LEEDS COLLEGE OF BUILDING
WITHDRAWN FROM STOCK

CIRIA Special Publication 96   1994

Environmental assessment

CONSTRUCTION INDUSTRY RESEARCH
AND INFORMATION ASSOCIATION
6 Storey's Gate
Westminster
London SW1P 3AU
Tel 071-222 8891
Fax 071-222 1708

THOMAS TELFORD SERVICES LTD
Thomas Telford House
1 Heron Quay
London E14 4JD
Tel: 071:987 6999
Fax: 071-538 4101

Sponsored by the Construction Sponsorship Directorate, Department of the Environment

Summary

This document sets out a basis for the Environmental Assessment (EA) of development schemes. Introductory chapters describe the stages involved in identifying and evaluating the potential impacts of projects, together with appropriate mitigation measures or enhancement opportunities. The document is applicable to all types and scales of development irrespective of whether a formal EA is required.

These principles are then developed further by considering specific types of development, although many of the principles will be relevant to other types of scheme. Separate chapters cover: Building development; River and coastal engineering; Linear development; Water and sewage schemes; Extractive industries; Waste management schemes; Electricity generation.

Each of the chapters draws on experiences with a number of projects, and, with the use of appropriate case studies, illustrates: the activities and features characteristic of such schemes; relevant planning and environmental legislation; the likely associated environmental impacts; means of evaluating these impacts; mitigation measures and enhancement opportunities.

The document is based on a conference held at Birmingham, UK in September 1992.

Environmental Assessment

Construction Industry Research and Information Association
Special Publication 96, 1994

Keywords:

Building, civil engineering, planning, transport, river and coastal engineering, waste management, minerals, electricity generation, water engineering, construction, environment, environmental assessment, handbook, good practice, design, conservation, environmental legislation.

Reader interest:

Clients, developers, engineers, planners, design and other consultants, Government, local authorities and regulatory bodies, statutory environmental consultees, academic and research organisations, environmental groups, legal and other advisers.

All rights reserved. No part of this publication may be reproduced or transmitted in any form or by any means, including photocopying and recording, without the written permission of the copyright holder, application for which should be addressesd to the publisher. Such written permission must also be obtained before any part of this publication is stored in a retrieval system of any nature.

© CIRIA 1994
ISBN 0 86017 379 8
Thomas Telford ISBN 0 7277 1999 8

	CLASSIFICATION
Availability	Unrestricted
Content	Guidance based on good practice
Status	Committee guided, Conference
User	Practitioners in planning, construction and environment

Published by CIRIA, 6 Storey's Gate, Westminster, London SW1P 3AU, in conjunction with Thomas Telford Services Ltd, Thomas Telford House, 1 Heron Quay, London E14 4JD.

Foreword

In 1989, as part of a review of its research activities, CIRIA's (then) Advisory Committee on Public Health and Hydraulic Engineering recommended that CIRIA intensified and consolidated its activities relating to environmental issues. This resulted in two initiatives. The first was to guide planners, designers and contractors on means of minimising the environmental impact of their development schemes. This led to CIRIA Research Project 424: Environmental Assessment, the results of which are the basis of this document.

This document presents an introduction to environmental assessment in Chapters 1 and 2, while Chapters 3–9 discuss a range of specific development types. Each of Chapters 3–9 was written by a separate author or authors to a common brief developed by a CIRIA-led working party in consultation with an industry-wide steering group. Each of the chapters has been subject to a separate review by a range of practitioners with particular knowledge and expertise of the type of development concerned.

An earlier edition of this document was presented at a seminar held in Birmingham in September 1992, and the document was subsequently modified to take account of feedback obtained at this event.

The second initiative followed in 1990 with a series of consultation meetings which led to the development and launch of the Construction Industry Environmental Forum. The initiative is operated by CIRIA, in partnership with the Building Research Establishment (BRE) and the Building Services Research and Information Association (BSRIA).

Acknowledgements

Authors

The production of this document has been undertaken largely on an in-kind basis. CIRIA expresses its gratitude for the personal commitment of the authors involved and to their respective organisations for providing the necessary in-kind resources for the study:

Michael Renger	-	Nabarro Nathanson
Jan Brooke	-	Posford Duvivier Environment
Carolyn Francis	-	Sir William Halcrow & Partners Ltd
Chris Ferrary	-	Environmental Resources Management (formerly ERL)
Tom Shaw	-	Shawater
Tony Robson, Colin Taylor	-	Nuclear Electric
Richard Clough	-	Wimpey Environmental
Jane Barron	-	W S Atkins Environment

Steering Group

CIRIA was guided by a Steering Committee drawn from many sectors of the construction industry. CIRIA expresses its gratitude to the following:

John Cowan	-	Resource Consultants, Cambridge
Jan Brooke	-	Posford Duvivier Environment
Tom Shaw	-	Shawater
Brian Morris	-	Laing
Richard Clough	-	Wimpey
Alan Bell	-	Cheshire County Council (representing RTPI)
Brian Saunders	-	British Railways Board
Graham Jukes	-	Institute of Environmental Health Officers
Jenny Heap	-	English Nature
Eric Potter	-	Council for the Protection of Rural England
Owen Jenkins	-	CIRIA

Particular thanks are expressed to John Cowan for Chairing the Committee and to Jan Brooke for advice on the development of the document.

Reviewers

An earlier draft of each of the chapters has been reviewed by a separate advisory panel. The contributions of those named below are gratefully acknowledged:

Building engineering
Graham Cully, English Nature
Tim Clark, University of York, IAAS
Shirley Watson, Grove Consultants
Brian Morris, Laing Technology
John Newton, Environmental Consultant

River and coastal engineering
Terry Oakes, Waveney District Council
Andrew Brookes, National Rivers Authority
F A Eric Hunter, MAFF/Consultant
Tony Edmonson, Harbour and General Works Ltd
Jenny Heap, English Nature

Linear development
John Smith, Travers Morgan Environment
Vivien Bodner, Department of Transport
Carol Wilson, British Rail
Chris Joel, Alfred McAlpine
A Ryder, RSK Environment
C Tuckwell, Cambridgeshire County Council
Jenny Heap, English Nature

Water engineering
M S Brown, South West Water
W Dunn, Biwater
John Cowan, Resource Consultants Cambridge
Alvin Smith, Binnie & Partners
Andrew Brookes, National Rivers Authority
Paul Hosé, English Nature

Mineral extraction
Mary Scott, DoE Minerals Division
Steve Fidgett, Sand and Gravel Association
Nick Coppin, Wardell Armstrong
Duncan Pollock, BACMI
John Kanesky, British Coal Opencast
Richard Kell, Leicester County Council

Waste management
Liz Chalk, National Rivers Authority
Chris Peters, London Waste Regulatory Authority
Alan Bell, Cheshire County Council
Christopher Murphy, Institute of Wastes Management

Electricity generation
Paul Johnson, Ove Arup & Partners
Mike Bailey, National Power
Russell Bowler, Ashdown Environmental
Fred Brown, London Borough of Greenwich
Tim Horwood, Cornwall County Council
P Quilleash, Sir Robert McAlpine/Renewable Energy Systems Ltd

Additional comments

CIRIA is also grateful to the following organisations who have commented on the document:

Association of County Councils
Association of District Councils
Association of Environmental Consultancies
Association of Metropolitan Authorities
Countryside Commission

County Surveyors Society
Department of the Environment Planning Directorate
District Planning Officers' Society
Her Majesty's Inspectorate of Pollution
Institute of Environmental Assessment

Authors' acknowledgements

The involvement of the following colleagues is gratefully acknowledged:

Sara Poulson	-	Nabarro Nathanson
Patrick Hawker	-	Sir William Halcrow & Partners Ltd
Tim Turner	-	Sir William Halcrow & Partners Ltd
Jill Rankin	-	Sir William Halcrow & Partners Ltd
Christine Hunter	-	Sir William Halcrow & Partners Ltd
Catriona Patterson	-	Posford Duvivier Environment
Cathy Thompson	-	Nuclear Electric
Ray Iredale	-	Noise Consultant
Stephen Hill	-	National Grid Company
Mary Mitchell	-	National Power
Joanne Kwan	-	CIRIA

Funders

CIRIA is most grateful to George Wimpey plc, John Laing plc, Thames Water and the Department of the Environment for contributing towards the cost of this study.

Contents

1	INTRODUCTION	1
	1.1 BACKGROUND	1
	1.2 PURPOSE AND SCOPE OF THIS DOCUMENT	1
	1.3 USE OF THIS DOCUMENT	2
	1.4 THE BROADER ENVIRONMENTAL PERSPECTIVE	3
	1.5 REFERENCES	5
2	APPROACH TO ENVIRONMENTAL ASSESSMENT	7
	2.1 LEGISLATION AND ADVISORY DOCUMENTS	7
	2.1.1 Aim of an EA	7
	2.1.2 Legislative framework	7
	2.1.3 Advisory documents	8
	2.1.4 Will an EA be required?	8
	2.2 INTRODUCTION TO EA	9
	2.2.1 Background	9
	2.2.2 Contents of an Environmental Statement	10
	2.2.3 Stages in an EA	10
	2.3 INITIAL CONSIDERATION OF DEVELOPMENT OPTIONS	12
	2.4 PRELIMINARY ASSESSMENT OF THE POSSIBLE ENVIRONMENTAL IMPACTS OF THE SCHEME	12
	2.4.1 Preliminary scoping exercise	12
	2.4.2 Preliminary consultation and data collection	15
	2.4.3 Types of impact	16
	2.4.4 Special sites	17
	2.5 ESTABLISHMENT OF THE ENVIRONMENTAL CHARACTERISTICS OF THE SITE	18
	2.5.1 General requirements	18
	2.5.2 Consultation	19
	2.5.3 Baseline data collection	21
	2.6 IDENTIFICATION OF POSSIBLE IMPACTS AND APPROPRIATE MITIGATION MEASURES	23
	2.6.1 Identification and prediction of impacts	23
	2.6.2 Mitigation measures	23
	2.6.3 Enhancement opportunities	26
	2.6.4 Contingency plans	27
	2.7 REFERENCES	28
3	BUILT DEVELOPMENT	29
	3.1 TYPE OF DEVELOPMENT CONSIDERED	31
	3.2 THE LEGAL FRAMEWORK	31

3.3	VOLUNTARY ENVIRONMENTAL IMPACT ASSESSMENTS	35
3.4	DETERMINING THE SCOPE OF THE ASSESSMENT	35
3.5	POTENTIAL ENVIRONMENTAL IMPACT OF BUILT DEVELOPMENT	37
	3.5.1 Land contamination	38
	3.5.2 Air pollution: dust and smoke	39
	3.5.3 Water pollution	40
	3.5.4 Nature conservation: ecology and earth sciences	40
	3.5.5 Landscape	42
	3.5.6 Noise	42
	3.5.7 Transport: traffic and congestion	43
	3.5.8 Public utilities	44
	3.5.9 Visual impact	45
	3.5.10 Socio-economic	45
	3.5.11 Heritage	46
	3.5.12 Other impacts	46
3.6	ENVIRONMENTAL ASSESSMENT IN PRACTICE: CASE STUDIES	47
	3.6.1 Case study 1	47
	3.6.2 Case study 2	48
3.7	REFERENCES	49
4	**RIVER AND COASTAL ENGINEERING**	**53**
4.1	TYPE OF SCHEME CONSIDERED	55
4.2	KEY STANDARDS AND LEGISLATION (1992)	55
	4.2.1 The need for Environmental Assessment	55
	4.2.2 Other relevant legislation	58
4.3	PRINCIPAL ACTIVITIES AND MAIN ENVIRONMENTAL EFFECTS	60
	4.3.1 The scoping exercise	60
	4.3.2 Baseline data requirements: project activities	61
	4.3.3 Common environmental impacts associated with river and coastal engineering works	63
	4.3.4 Baseline data requirements: environmental parameters	66
	4.3.5 The role of consultation in the data-collection process	66
	4.3.6 Supplementary data collection	68
4.4	IMPACT IDENTIFICATION, EVALUATION AND MITIGATION	70
	4.4.1 Impact assessment process	70
	4.4.2 Water quality, water resources and groundwater	71
	4.4.3 Sediment quality, soil and geology	73
	4.4.4 Physical processes: hydrology, geomorphology, hydraulic regime, coastal configuration	74
	4.4.5 Air quality	76
	4.4.6 Natural habitats	76
	4.4.7 Socio-economic and local community	78
	4.4.8 Leisure and amenity/tourism	79
	4.4.9 Land use and landscape	81
	4.4.10 Health and safety; emergency services; risk	82
	4.4.11 Traffic: transport, navigation and infrastructure	84
	4.4.12 Heritage and archaeology	84
4.5	POST-PROJECT ASSESSMENT REQUIREMENTS	84

	4.6	REFERENCES AND BIBLIOGRAPHY	85

5 WATER SUPPLY INFRASTRUCTURE AND WASTEWATER TREATMENT WORKS — 89

    5.1 TYPE OF SCHEME CONSIDERED — 91

    5.2 KEY STANDARDS AND LEGISLATION — 92
        5.2.1 Legislation governing Environmental Assessments — 92
        5.2.2 Legislation on conservation issues — 92
        5.2.3 Legislation on water supply and pollution control — 96
        5.2.4 Legislation concerning construction in the water industry — 101

    5.3 PRINCIPAL EA AND CONSTRUCTION ACTIVITIES — 102
        5.3.1 Environmental Assessment methodology — 102
        5.3.2 Summary of main engineering activities: water supply projects — 106
        5.3.3 Summary of main engineering activities: wastewater treatment works — 111

    5.4 IMPACT IDENTIFICATION, EVALUATION AND MITIGATION — 114
        5.4.1 Air quality — 114
        5.4.2 Water quality — 117
        5.4.3 Water resources — 119
        5.4.4 Land — 120
        5.4.5 Physical processes — 122
        5.4.6 Impact on natural habitats — 123
        5.4.7 Community structure — 126
        5.4.8 Health and safety — 126
        5.4.9 Socio-economics — 128
        5.4.10 Leisure and amenity — 128
        5.4.11 Land use and landscape — 128
        5.4.12 Noise — 130
        5.4.13 Traffic and navigation — 130
        5.4.14 Cultural heritage — 130

    5.5 CONCLUSIONS — 130

    5.6 REFERENCES — 131

6 LINEAR DEVELOPMENT — 133

    6.1 TYPE OF SCHEME CONSIDERED — 135

    6.2 KEY STANDARDS, LEGISLATION AND ADVICE — 135
        6.2.1 Introduction — 135
        6.2.2 Requirements for EA — 136
        6.2.3 Government advice on EA — 137
        6.2.4 Other advisory documents — 138

    6.3 PRINCIPAL ACTIVITIES IN THE ASSESSMENT — 138
        6.3.1 Introduction — 138
        6.3.2 The scoping exercise — 139
        6.3.3 The project description — 139
        6.3.4 Baseline data requirements — 140
        6.3.5 Consultations — 141
        6.3.6 Presentation — 141
        6.3.7 After the ES — 142

6.4	IMPACT IDENTIFICATION, EVALUATION AND MITIGATION	142
	6.4.1 Introduction	142
	6.4.2 Physical and chemical effects: vibration	143
	6.4.3 Physical and chemical effects: air quality	144
	6.4.4 Physical and chemical effects: water quality	147
	6.4.5 Natural habitats: effects on ecology	150
	6.4.6 Human characteristics: noise	151
	6.4.7 Human characteristics: severance and other traffic effects	155
	6.4.8 Human characteristics: visual impacts	156
	6.4.9 Human characteristics: effects on archaeology and historic buildings	157
	6.4.10 Socio-economic effects	159
6.5	REFERENCES	162

7 ELECTRICITY GENERATION — 165

7.1	INTRODUCTION	167
7.2	KEY LEGISLATION ON EAs FOR ELECTRICITY GENERATION	168
	7.2.1 Legal framework	168
	7.2.2 EA process and documentation	170
	7.2.3 Environmental Statement	172
	7.2.4 Supporting reports	173
	7.2.5 Non-technical summary	174
7.3	INTEGRATED POLLUTION CONTROL (IPC)	174
7.4	AIR QUALITY	175
	7.4.1 Existing environment	175
	7.4.2 Air quality regulations and standards	177
	7.4.3 Impact evaluation	178
	7.4.4 Mitigation measures	182
7.5	WATER QUALITY	183
	7.5.1 General	183
	7.5.2 Water quality regulations and standards	183
	7.5.3 Aquatic impact evaluation	187
	7.5.4 Mitigation	191
7.6	SOLID WASTES	191
	7.6.1 Existing environment	191
	7.6.2 Solid waste quality standards, regulations and guidelines	191
	7.6.3 Impact evaluation	192
	7.6.4 Mitigation	196
7.7	NOISE AND VIBRATION	197
	7.7.1 General and noise background	197
	7.7.2 Noise and vibration regulations and standards	197
	7.7.3 Noise impacts	198
	7.7.4 Vibration	201
7.8	VISUAL IMPACTS	201
	7.8.1 General	201
	7.8.2 Impact evaluation	201
	7.8.3 Mitigation	202

7.9	INTERFERENCE WITH RADIO COMMUNICATIONS	202
	7.9.1 Wind turbines	202
	7.9.2 Transmission lines	203
7.10	ELECTROMAGNETIC FIELDS	203
	7.10.1 General	203
	7.10.2 Regulation and standards	203
	7.10.3 Impact evaluation	203
7.11	HERITAGE FEATURES (ENGLAND AND WALES)	203
7.12	ECOLOGICAL APPRAISAL: TERRESTRIAL	204
	7.12.1 General	204
	7.12.2 Impact evaluation	205
	7.12.3 Mitigation	210
7.13	OTHER SOURCES	211
	7.13.1 Biomass crops for energy	211
	7.13.2 Photovoltaics	211
	7.13.3 Landfill gas	211
	7.13.4 Refuse-derived fuel	211
	7.13.5 Geothermal	213
	7.13.6 Hydro power	213
	7.13.7 Tidal power	213
	7.13.8 Sea wave power	213
7.14	CONCLUSIONS	214
7.15	REFERENCES	214

8	MINERALS EXTRACTION	219
8.1	TYPE OF SCHEME CONSIDERED	221
8.2	KEY STANDARDS, LEGISLATION AND SUPPORTING ADVICE	222
	8.2.1 Environmental Assessment requirements	222
	8.2.2 Legislation affecting minerals extraction	222
	8.2.3 Control of existing environmental impacts	227
	8.2.4 Other documents on minerals planning	228
8.3	PRINCIPAL ACTIVITIES AND MAIN ENVIRONMENTAL EFFECTS	230
	8.3.1 The scoping exercise	230
	8.3.2 Baseline data requirements: project activities	230
	8.3.3 Common environmental impacts associated with minerals extraction	232
	8.3.4 Baseline data requirements: environmental parameters	232
	8.3.5 The role of consultation in the data-collection process	235
8.4	IMPACT, IDENTIFICATION, EVALUATION AND MITIGATION	235
	8.4.1 Human impacts: general	236
	8.4.2 Noise	236
	8.4.3 Vibration	239
	8.4.4 Visual impacts	240
	8.4.5 Traffic and transport	240
	8.4.6 Land use	241
	8.4.7 Heritage and archaeology	242
	8.4.8 Air quality	242
	8.4.9 Hydrogeology	244
	8.4.10 Surface water	245
	8.4.11 Geology and geomorphology	245

		8.4.12 Natural habitats	246
	8.5	REFERENCES	246
9	WASTE MANAGEMENT		249
	9.1	TYPES OF SCHEME CONSIDERED	251
		9.1.1 Waste arisings	251
		9.1.2 Waste management methods and environmental protection	251
	9.2	KEY STANDARDS AND LEGISLATION	254
		9.2.1 Environmental Assessment requirements	254
		9.2.2 Other relevant legislation	254
		9.2.3 Responsibilities of controlling agency	255
		9.2.4 Public perception	256
	9.3	PRINCIPAL ACTIVITIES AND MAIN ENVIRONMENTAL EFFECTS	260
		9.3.1 The scoping exercise	260
		9.3.2 Baseline data requirements: project activities	260
		9.3.3 Common environmental impacts associated with waste management	261
		9.3.4 Baseline data requirements: environmental parameters	262
		9.3.5 The role of consultation in the data-collection process	263
	9 4	IMPACT IDENTIFICATION, EVALUATION AND MITIGATION	267
		9.4.1 General	267
		9.4.2 Air quality	267
		9.4.3 Water: water quality and groundwater	269
		9.4.4 Land: soil and geology	270
		9.4.5 Flora and fauna	271
		9.4.6 Socio-economic/cultural	273
		9.4.7 Leisure and amenity/tourism	274
		9.4.8 Land use and landscape	274
		9.4.9 Health and safety; emergency services; risk	275
		9.4.10 Traffic: transport and infrastructure	276
		9.4.11 Heritage and archaeology	278
		9.4.12 Noise	278

List of Figures

Figure 2.1 *Establishing requirement for formal EA*	9
Figure 4.1 *The environmental assessment process*	64
Figure 4.2 *Environmental impacts of proposed oil berth development*	65
Figure 6.1 *Summary of phases and components of light rail schemes likely to give rise to impacts*	140
Figure 7.1 *Frequency distribution of SO_2 concentrations from 2000 MW power station at radius of maximum effect (~8 km)*	179
Figure 7.2 *Peak hourly average GLCs*	179
Figure 7.3 *Estimated annual average ground level concentrations of NO_x ($NO + NO_2$) CCGT power station site*	179
Figure 7.4 *Wind turbine and wind-generated background noise versus windspeed (turbine design and location specific)*	201
Figure 7.5 *Habitat types in survey area*	206
Figure 7.6 *Plant species in survey area*	207
Figure 8.1 *The environmental assessment process (after Brooke and Whittle, 1990)*	224
Figure 8.2 *Minerals planning applications*	225
Figure 8.3 *Typical environmental impacts of extractive processes*	233

Figure 8.4 *Attitudes of people living near minerals extractions to their environmental effects*   237

List of Boxes

Box 1.1	Broader environmental issues: Environmental issues in construction	4
Box 2.1	Examples of scheme-specific EA regulations	7
Box 2.2	Requirements of the regulations as to the content of Environmental Statements	11
Box 2.3	Case study: Scoping	13
Box 2.4	Project life-cycle	14
Box 2.5	Activities associated with construction projects	15
Box 2.6	Nature conservation and additional designations	18
Box 2.7	Examples of data subject to seasonal or other variations	22
Box 2.8	Case study: Mitigation	24
Box 2.9	Examples of mitigation measures	25
Box 2.10	Case study: Enhancement	27
Box 3.1	Source documents	32
Box 3.2	Advisory thresholds for EA	33
Box 3.3	Factors determining desirability of a voluntary EA	35
Box 3.4	Potential dust sources	39
Box 3.5	Potential impacts: matters to consider	39
Box 3.6	Statutory controls over construction noise	42
Box 3.7	Sources of noise	43
Box 3.8	Impact on highway network material considerations	44
Box 3.9	Socio-economic effects	46
Box 4.1	Categorisation of impacts arising from river and coastal engineering works	71
Box 4.2	Case study: Hydrological mitigation	74
Box 4.3	Case study: Physical process impacts	74
Box 4.4	Case study: Habitat protection	78
Box 4.5	Case study: Recreation benefits	81
Box 4.6	Case study: Landscaping opportunities	82
Box 5.1	Bellozanne sludge dryer, Jersey	115
Box 5.2	Water quality modelling	117
Box 5.3	Thames chalk streams low flow study	119
Box 5.4	Roadford reservoir scheme	120
Box 5.5	Environmentally sensitive operating rules for the Roadford scheme	125
Box 7.1	EA process (construction and operation)	171
Box 7.2	Exposure to radiological discharges	174
Box 7.3	Some air quality issues	175
Box 7.4	Frequency and spatial distribution of pollutants	179
Box 7.5	Example of specific noise limits	199
Box 7.6	Archaeological and listed building surveys (example)	204
Box 7.7	Key environmental issues for selected electricity generation technologies	212
Box 8.1	Criteria for Environmental Assessments in the minerals industry	223
Box 8.2	Official publications on minerals Environmental Assessment	226
Box 8.3	Department of the Environment Circulars on planning	229
Box 8.4	Case study: Regrading of land	241
Box 8.5	Environmental Protection Act guidance	243
Box 9.1	Waste reduction at source: reuse and recycling	252
Box 9.2	Trends and specific problem areas	253
Box 9.3	Case study: The Welbeck Project	264
Box 9.4	Case study: Witton Landfill	272
Box 9.5	Case study: Loscoe	276
Box 9.6	Case study: Stone landfill gas plant	276
Box 9.7	Case study: Arpley bridge, rail terminal now being considered	278

List of Tables

Table 2.1	Examples of physical and chemical effects	16
Table 2.2	Possible consultees	20
Table 3.1	Other statutory sources	34
Table 3.2	Potential consultees	37
Table 3.3	Factors that may influence the significance of an impact	38
Table 4.1	Examples of river and coastal engineering projects	56
Table 4.2	Summary of UK legislation relevant to the EA process for river and coastal engineering	59
Table 4.3	International legislation relevant to the EA process for river and coastal engineering	59
Table 4.4	Summary of European Directives relevant to the EA process for river and coastal engineering	60
Table 4.5	Activities that may result in potentially significant impacts	62
Table 4.6	Baseline data requirements	67
Table 4.7	Possible consultees in the EA process	69
Table 4.8	Examples of the impact of river and coastal engineering works on water quality and water resources	72
Table 4.9	Examples of the impact of river and coastal engineering works on soil, sediments and geological interest	73
Table 4.10	Examples of the impact of river and coastal engineering works on physical processes	75
Table 4.11	Examples of the impact of river and coastal engineering works on natural habitats	77
Table 4.12	Examples of the impact of river and coastal engineering works on socio-economic parameters	79
Table 4.13	Examples of the impact of river and coastal engineering works on leisure and amenity	80
Table 4.14	Examples of the impact of river and coastal engineering works on land use and landscape	83
Table 4.15	Examples of the impact of river and coastal engineering works on safety characteristics	83
Table 5.1	Examples of water supply infrastructure and wastewater treatment schemes	91
Table 5.2	Summary of UK legislation affecting water supply and wastewater treatment schemes	94
Table 5.3	Summary of European Community Council Directives affecting water supply infrastructure and wastewater treatment schemes	98
Table 5.4	International Conventions on conservation	99
Table 5.5	Baseline data requirements	104
Table 5.6	Possible consultees in the EA process in the UK	105
Table 5.7a	Activities that may result in potentially significant impacts (construction phase)	108
Table 5.7b	Activities that may result in potentially significant impacts (operational and post-operational phases)	109
Table 5.8	Summary of impacts considered	114
Table 5.9	Examples of the impact of water supply and wastewater treatment works on air quality	116
Table 5.10	Examples of the impact of water supply and wastewater treatment works on water quality and water resources	118
Table 5.11	Examples of the impact of water supply and wastewater treatment works on land	121
Table 5.12	Examples of the impact of water supply and wastewater treatment works on physical processes	122
Table 5.13	Examples of the impact of water supply and wastewater treatment works on natural habitats	124
Table 5.14	Examples of the impact of water supply and wastewater treatment works on health and safety characteristics	127

Table 5.15	Examples of the impact of water supply and wastewater treatment works on socio-economic parameters	127
Table 5.16	Examples of the impact of wastewater supply and waste treatment works on leisure and amenity	129
Table 5.17	Examples of the impact of water supply and wastewater treatment works on land use, landscape, and cultural heritage	129
Table 6.1	Examples of linear development projects	135
Table 6.2	Categorisation of impacts arising from linear developments	143
Table 6.3	Examples of vibration impacts from linear developments	143
Table 6.4	Examples of the impact of linear development schemes on air quality	145
Table 6.5	Air quality standards	147
Table 6.6	Examples of the impact of linear development schemes on air quality	147
Table 6.7	Examples of the impact of linear development schemes on ecology	150
Table 6.8	Examples of the impact of linear development schemes on noise	152
Table 6.9	Criteria for evaluating the significance of noise during construction (external levels)	153
Table 6.10	Examples of traffic effects associated with linear developments	155
Table 6.11	Examples of visual effects associated with linear developments	156
Table 6.12	Examples of impact of linear development schemes on archaeology and historic buildings	157
Table 6.13	Examples of socio-economic effects of linear developments	159
Table 7.1	Possible consultees in the EA process (indicative only)	170
Table 7.2	Examples of ES chapter headings	172
Table 7.3	Aspects associated with potentially significant impacts (broad indication only)	173
Table 7.4	Air quality monitoring sites: December 1990	176
Table 7.5	EC and WHO air quality limit and guideline values (in $\mu g/m^3$, standardised at 293 K and 101.3 kPa pressure)	177
Table 7.6	Typical emission rates for 3×660 MW coal-fired station (without FGD)	178
Table 7.7	Summary of best estimate airborne discharges for a PWR reactor	181
Table 7.8	Examples of exposure levels	181
Table 7.9	Estimated maximum individual risk of death consequent upon PWR operation	181
Table 7.10	Summary of EC Directives relevant to EA process for energy schemes	185
Table 7.11	Proposed (UK) use-related water quality standards	186
Table 7.12	Effect of treated FGD effluent on a typical power station cooling water discharge	188
Table 7.13	Comparison of published commercial fish landings for ICES sub-area Vllf (Bristol Channel) with annual screen catches from Hinkley Point A and B power stations estimated from monthly samples	188
Table 7.14	Summary of liquid discharges for a PWR reactor (case study)	190
Table 7.15	Dietary and behavioural factors for critical groups identified for liquid effluent discharge assessments (case study)	190
Table 7.16	Estimated major raw material requirement and solid product arising during the operation of a 2×900 MW power station	194
Table 7.17	Classification of radioactive waste	194
Table 7.18	Estimated arisings of operational waste at an AGR reactor	195
Table 7.19	Composition of sludge from typical FGD effluent treatment	196
Table 7.20	Sources of existing ecological information	205
Table 7.21	Nature conservation designation: main types	208
Table 7.22	Examples of major habitat types (case study)	209
Table 7.23	Examples of impacts of power stations on ecology	210
Table 8.1	Examples of types of surface mineral extraction for which planning permission will be sought	222
Table 8.2	Activities that may result in potentially significant impacts	231
Table 8.3	Baseline data requirements	234
Table 8.4	Possible consultees in the EA process	235
Table 8.5	Summary of impacts considered	236
Table 8.6	Examples of the impact of minerals extraction on noise levels	238
Table 8.7	Examples of the impact of minerals extraction from blasting and vibration	239

Table 8.8	Examples of the visual impact of minerals extraction	240
Table 8.9	Examples of the impact of minerals extraction on traffic and transport	240
Table 8.10	Examples of the impact of minerals extraction on land use	241
Table 8.11	Examples of the impact of minerals extraction on heritage and archaeology	242
Table 8.12	Examples of the impact of minerals extraction on air quality	242
Table 8.13	Examples of the impact of minerals extraction on hydrogeology	244
Table 8.14	Examples of the impact of minerals extraction on surface water	245
Table 8.15	Examples of the impact of minerals extraction on geology and geomorphology	245
Table 8.16	Examples of the impact of minerals extraction on natural habitat	246
Table 9.1	Examples of waste management operations	253
Table 9.2	Summary of UK legislation relevant to the EA process	257
Table 9.3	Summary of EC Directives relevant to the EA process	258
Table 9.4	Waste Management Papers	259
Table 9.5	Activities that may result in potentially significant impacts	262
Table 9.6	Baseline data requirements	263
Table 9.7	Summary of the impacts considered in this chapter	267
Table 9.8	Examples of the impact of waste management schemes on air quality	268
Table 9.9	Examples of the impact of waste management schemes on water resources	269
Table 9.10	Examples of the impact of waste management schemes on flora and fauna	273
Table 9.11	Examples of the impact of waste management schemes on land use and landscape	275

1 Introduction

1.1 BACKGROUND

The consideration of the environmental implications of development schemes is by no means new. The term Environmental Impact Assessment is one that is familiar to most involved in construction. However, traditionally, the term and indeed the process have been associated with large-scale schemes of regional significance.

Certain types of development in the UK have also been the subject of detailed scrutiny at the planning stage for some time. It was not until the implementation in 1988 of the European Community (EC) Directive on 'The assessment of the effects of certain public and private projects on the environment' (85/337/EEC), and its implementation in the UK through various Statutory Instruments, including the Town and Country (Assessment of Environmental Effects) Regulations 1988 (SI No 1199) however, that the procedures became formalised in the UK. The process is now generally referred to as Environmental Assessment (EA).

At the same time, increasing environmental awareness from Government, industry and the public has led to a greater consideration of environmental issues in many aspects of political, commercial and domestic life.

Owing to the nature of their activities, the construction and related industries have been the subject of much scrutiny on environmental grounds. There is growing concern among construction practitioners at all levels to ensure that their activities, services and products lead to minimal adverse environmental impact.

Furthermore, while there has been much evidence of polarisation in the views held by the industry and environmental interest groups, there is increasing evidence of positive collaboration in finding solutions to environmental problems.

1.2 PURPOSE AND SCOPE OF THIS DOCUMENT

The purpose of this document is to heighten the awareness of both construction and environmental groups of the interaction between development schemes, their related activities and the environment. It encourages consideration of possible environmental issues from the initial planning stages of a scheme, through to the design, construction and operational stages, and to arrangements for closure/decommissioning where this is relevant.

The document recognises that, while most building and construction activities will lead to long-term benefits to the community and society as a whole, the location of these developments, and the way they are planned, designed, constructed and operated can have environmental implications. It therefore provides information on the engineering and operational activities associated with a range of different development schemes, together with their likely environmental effects. Guidance is also provided on appropriate measures available to identify the nature and extent of these impacts and, where potentially adverse impacts are identified, on the measures likely to avoid or minimise their impact with reference to case studies. Where appropriate, the document also describes opportunities for environmental enhancement.

The document is applicable to developments of all scales, irrespective of whether formal Environmental Assessment is required, and is aimed at a broad readership including Government, planning authorities, developers, design and other consultants, and environmental interest groups.

By discussing a range of schemes in terms of both the engineering and the environmental perspectives, it is believed that the document will encourage and facilitate dialogue between clients, designers and environmental interest groups, assisting in the promotion and development of schemes with due regard to their environmental effects, while acknowledging the need for flexibility in confronting uncertainties and conflicts.

Following a general introduction in Chapter 1, Chapter 2 introduces the concept of environmental assessment, describing the stages, parties and processes involved. It also outlines relevant Statutory Instruments, along with the mechanisms for determining whether a formal EA is required.

Chapter 3 develops the general principles introduced in Chapters 1 and 2, applying them to building development schemes. Although it relates primarily to residential, commercial, retail and leisure development, much of its contents is, however, relevant to other types of scheme. The remainder of the document consists of a number of chapters, each of which considers a particular type of development scheme. Each of these chapters adopts a similar format, and describes:

- the activities and features characteristic of such schemes
- relevant planning and environmental legislation
- the likely associated environmental impacts
- means of evaluating these impacts
- mitigation measures and enhancement opportunities.

Six specific types of development scheme are considered, reflecting a wide range of development and construction schemes:

- linear development
- water and sewage schemes
- river and coastal engineering works
- extractive industries
- waste management schemes
- electricity generation.

However, the information presented will have relevance to many other types of project.

The document is not intended to set prescriptive standards as to what should constitute an EA, nor does it offer exhaustive guidance on specific aspects of data collection and impact assessment. The reader is referred to other specialised documents such as those produced by English Nature and the National Rivers Authority (English Nature, 1994; National Rivers Authority, 1993).

1.3   USE OF THIS DOCUMENT

This document is aimed at a broad readership, and its structure and content reflect the differing requirements of this audience.

General interest

Readers interested in obtaining an overview of the process of EA (whether formal or otherwise), together with a broad appreciation of some of the common problems associated with construction, will find Chapters 1, 2 and 3 of particular interest.

Specific scheme-related interest

Readers wishing to gain an appreciation of the activities and issues involved with a particular type of development may proceed directly to the appropriate chapter (4–9) where a detailed discussion will be found. However, Chapters 1 and 2 may provide additional background information of relevance. Chapters 4–9 are largely self-contained. This facilitates the investigation by readers of specific types of scheme, but at the expense of some repetition between sections.

Specific issue-related interest

The standard format adopted in Chapters 3–9 permits the reader to identify information relating to specific environmental issues – pollution, impacts on humans such as noise, nature conservation issues etc.

1.4 THE BROADER ENVIRONMENTAL PERSPECTIVE

The regulatory framework that implements Directive 85/337/EEC provides a reasonable degree of control in respect of the environmental impact of development. This is complemented for smaller schemes by planning and other controls. However, this framework alone cannot ensure that the overall environmental impact of the development, both in terms of its construction and its operation, is minimised.

There are many environmental issues that will not always be addressed by EA. Examples include the energy efficiency of buildings and appliances and the source of materials used in construction.

It is likely that the attention given to the life cycle environmental impact of developments – from initial planning considerations through to materials production, construction and eventual operation and demolition – will continue to increase. Environmental criteria are now being used to judge services and products, and the provision of environmental information on product labels is now widely practised. Labelling or rating schemes are also available that consider the energy efficiency of buildings, and with the introduction of the Building Research Establishment's Environmental Assessment Method (BREEAM) (BRE 1990, 91, 92, 93), mechanisms exist to judge the environmental performance of entire buildings. While these types of issues are likely to become increasingly important for construction in its pursuit of sustainable development, further discussion of this concept is beyond the scope of this document.

The broader environmental issues affecting construction are considered in CIRIA's report, *Environmental issues in construction* (CIRIA, 1993a), which considers all aspects relating from the winning and production of materials through to the operation, refurbishment and demolition of buildings and structures.

The report considers environmental impacts in terms of:

- the use of energy
- the use of resources
- the creation of pollution
- specific operational aspects (particularly those relating to the internal environment)
- land use, planning and conservation measures.

The interrelationships between the above are also an important consideration.

Box 1.1 adapted from CIRIA (1993a) describes the broader implications of these issues. These are also being considered in CIRIA project 459 (CIRIA, 1993b) – *Environmental checklists for building and construction.*

Box 1.1 Broader environmental issues: Environmental issues in construction

Energy use, global warming and climate change
The use of fossil-fuel-derived energy in the production of materials, during the construction process, and by the occupants or users of the building or structure throughout its lifetime, is a source of significant quantities of carbon dioxide, which is considered to contribute to global warming. The energy used in buildings accounts for approximately 50% of the total used in the UK. The design of the buildings, in terms both of the thermal performance of their fabric and the appliances and services used within them, can therefore be a significant factor in minimising environmental impact.

Resources, waste and recycling
The construction industry is a conspicuous user of resources. Materials are derived from numerous sources and suppliers, and minimisation of waste presents particular problems. Although many of the materials in use are common to most sites, the fragmented nature of development at present constrains the practical extent of recycling. The use of materials can also create numerous secondary impacts in terms of disturbance to habitats and pollution associated with manufacturing processes and transport.

Pollution and hazardous substances
The operational pollution effects of a development, whether sewage from a housing development or stack emissions from an industrial process, will be considered as part of the EA process. However, the design and construction phases involve the specification of materials, and the use of plant, processes and techniques. Construction also involves extensive disturbances to the existing environment, whether on green field or previously developed sites. Each of these activities poses a risk of introducing pollutants into the environment that can affect the workers on site, the neighbourhood, or the local ground, water and air quality. In many cases pollution effects can be regional or global. Similar impacts can occur during the operational phase of the development.

Internal environment
In the developed world, human beings spend up to 90% of their lives within buildings. They are exposed to a range of chemicals arising from furnishing, finishes and other practices which take place within the building. Increasingly, the design and layout of buildings necessitate active measures to maintain conditions that ensure the health and general well-being of their occupants.

Land use, planning and conservation
The location of developments initiated by the client and built by the construction industry is largely controlled by formal regulatory procedures. However, the success of the development in integrating with, and the acceptability of the way in which it modifies and interacts with the surrounding natural and built environment, cannot be ensured wholly by regulations.

Effective land-use planning is fundamental to integrating conservation and development. It should incorporate the principles of environmentally led planning and sustainable development. This means that land-use planning should be underpinned by environmental considerations and should be carried out in such a way as to make clear to developers and other users of the land the views of local authorities and local communities about particular development proposals in particular areas.

1.5 REFERENCES

CIRIA (1993a)
Environmental issues in construction
CIRIA (London)

CIRIA (1993b)
Environmental handbooks for building and civil engineering
Volume 1 Design and specification
Volume 2 Construction phase
CIRIA (London)

BRE (1990, 91, 92, 93)
Building Research Establishment Environmental Assessment Method (BREEAM)
Building Research Establishment (Watford)

EEC (1985)
The assessment of the effects of certain public and private projects on the environment
(85/337/EEC)

English Nature (1994)
Nature Conservation in Environmental Assessment
English Nature (Peterborough)

NRA (in press)
Assessing Environmental Impact in the NRA — A Technical Guide to EA Function Applicability
NRA (Bristol)

2 Approach to Environmental Assessment

2.1 LEGISLATION AND ADVISORY DOCUMENTS

This section discusses the regulatory framework associated with formal Environmental Assessment (EA). Discussion is limited to that applicable in England and Wales although some reference is made to legislation in Scotland and Northern Ireland. It also makes reference to material of relevance to development schemes that would not require formal EA. The technical breadth of this document makes it necessary to cover an extensive range of legislation. Whilst every effort has been made to ensure the correctness of the information presented in the document at the time of writing (September 1992), readers should check for recent changes.

2.1.1 Aim of an EA

Although it has been the practice of developers for many years to investigate the environmental effects of large-scale projects as part of the normal project-planning process, the legal requirements to do so date from as recently as 1988. The progenitor of the present legal requirements is the European Community Directive No 85/337. This Directive sets out two categories of project, one for which EA must always be carried out (listed in Annex I) and others for which EA is required if in the opinion of the competent authority the project is likely to have significant environmental effects (listed in Annex II). The Directive also specifies the topics that should be covered by information supplied by the developer when EA is required; in UK implementing legislation this is known as the Environmental Statement (ES). One particular facet of the EA is to establish the nature of the existing characteristics and conditions of the site and its locality.

2.1.2 Legislative framework

In the UK, the Directive has been implemented through a number of regulations, many of which have been issued under the provisions of various existing Acts. Most building development projects requiring EA would come under the remit of the Town and Country Planning (Assessment of Environmental Effects) Regulations 1988. However, several other regulations exist that relate to specific types of development. Some of the regulations relevant to the schemes covered in this document are stated in Box 2.1.

Box 2.1 Examples of scheme-specific EA regulations

Town and Country Planning (Assessment of Environmental Effects) Regulations 1988

Land Drainage Improvements Works (Assessment of Environmental Effects) Regulations 1988

Harbour Works (Assessment of Environmental Effects) Regulations 1988

Highways (Assessment of Environmental Effects) Regulations 1988

Harbour Works (Assessment of Environmental Effects) (No. 2) Regulations 1989

The Electricity and Pipe-line Works (Assessment of Environmental Effects) Regulations 1990

Similar instruments exist in other European Countries.

2.1.3 Advisory documents

Guidance on the aforementioned Directive and Regulations has been produced by the UK Government. DoE Circular 15/88 (DoE/Welsh Office, 1988), *Environmental Assessment*, gives guidance on interpretation of the Regulations, while the HMSO publication *Environmental Assessment — A guide to the procedures* (HMSO, 1989) also contains much guidance for practitioners (also known as the Blue Book).

2.1.4 Will an EA be required?

Early consideration of the environmental issues relating to a development scheme, and anticipation of any impacts, will often avoid a good deal of delay and possible expense, irrespective of whether the project is of a scale or significance likely to require formal EA. However, it is clearly in the developer's interest to establish at the earliest stage in the project planning whether a formal EA is likely to be required. For certain large-scale projects, initial planning will inevitably involve a wide-ranging review of alternative proposals for the site and the establishment of potential land-use options. The nature of these proposals and the nature of the site will both determine whether an EA is necessary. Indicative criteria and thresholds for identifying projects requiring EA are provided in DoE Circular 15/88 (DoE/Welsh Office, 1988). These are also covered for specific schemes in later chapters. Often, the most effective way to resolve the issue is to ask the appropriate regulatory authority at the outset whether or not they will require an EA. If the authority replies in the affirmative, then the developer may appeal to the Secretary of State under Section 5(6) of the Regulations (see Figure 2.1).

A development listed in Schedule 2 will only require an EA if it is likely to have 'significant effect' on the environment. Most schemes will have an effect on the environment but not all effects will be significant. DoE Circular 15/88 suggests that there are three main criteria for significance:

- whether the project is of more than local importance, principally in terms of physical scale

- whether the project is intended for a particularly sensitive location, for example, a National Park or a Site of Special Scientific Interest, and for that reason may have a significant effect on the area's environment even though the project is not on a major scale. Additional advice is provided in Circular 1/92, *Planning Controls over Sites of Special Scientific Interest* (DoE 1992a)

- whether it is thought that the project is likely to give rise to particularly complex or adverse effects, for example in terms of discharge of pollutants.

However, even if a formal EA is not required, adopting a rational framework for the collection of relevant environmental information to the extent of submitting a voluntary environmental statement will help to identify possible impacts and allow appropriate mitigation measures to be adopted. It will help to establish dialogue with interest groups, and facilitate early identification of possible issues of concern, thereby reducing the risk of unforeseen problems and consequent costly delays.

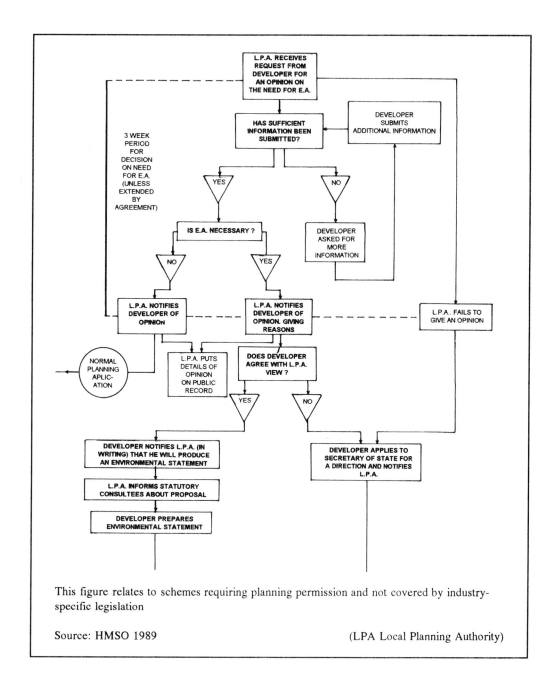

Figure 2.1   *Establishing requirement for formal EA*

2.2 INTRODUCTION TO EA

2.2.1 Background

An EA is a systematic analysis, using the best practicable techniques and best available sources of information, of the environmental effects of a project. It is initiated and prepared by the developer of the scheme, usually in consultation with others, and the results of the assessment are considered by the planning or other statutory authority in forming its judgement as to whether, on environmental grounds, the development should go ahead.

General guidance on the execution of an environmental assessment and the contents of the resulting report — the Environmental Statement — are provided in the DoE 'Blue Book' (HMSO, 1989) (see also Box 2.2), and Circular 15/88.

This chapter outlines the general approach to the initiation, planning and execution of an environmental assessment. The framework described is applicable to all schemes irrespective of scale and whether a formal EA is required, and underlines:

- the importance of an interactive and multidisciplinary approach (see Section 2.4.1)
- the importance and benefits of early and continuing consultation
- the iterative nature of the process.

2.2.2 Contents of an Environmental Statement

The environmental assessment process should be undertaken with a clear view of the purpose and required contents of the Environmental Statement. The statement should include the following components:

- a description of the development proposed
- data identifying and assessing the effects of the development on the environment
- a description of the significant environmental effects
- a description of the measures to be taken to avoid, reduce or remedy any adverse effects
- a summary in non-technical language.

The contents of an Environmental Statement are described in Box 2.2.

2.2.3 Stages in an EA

The environmental assessment process is shown in Figure 4.1, which illustrates its interactive and iterative nature. The stages involved include:

- initial consideration of development options
- preliminary assessment of the possible environmental impacts of the scheme
- establishment of the environmental characteristics of the site
- identification and assessment of environmental impacts
- identification of potential mitigation measures.

The stages are interrelated and will involve consultation, data collection and analysis, the extent of which will clearly be influenced by the nature of the development and the characteristics of the site.

The remainder of this chapter has been written to reflect these stages.

Box 2.2 Requirements of the regulations as to the content of Environmental Statements

The following are the statutory provisions with respect to the content of Environmental Statements, as set out in Schedule 3 to the Town and Country Planning (Assessment of Environmental Effects) Regulations 1988:

1. An Environmental Statement comprises a document or series of documents providing for the purpose of assessing the likely impact upon the environment of the development proposed to be carried out, the information specific in paragraph 2 (referred to in this Schedule as 'the specified information').

2. The specified information is:
   (a) a description of the development proposed, comprising information about the site and the design and size or scale of the development;

   (b) the data necessary to identify and assess the main effects which that development is likely to have on the environment;

   (c) a description of the likely significant effects, direct and indirect, on the environment of the development, explained by reference to its possible impact on:
   human beings;
   flora;
   fauna;
   soil;
   water;
   air;
   climate;
   the landscape;
   the interaction between any of the foregoing;
   material assets;
   the cultural heritage;

   (d) where significant adverse effects are identified with respect to any of the foregoing, a description of the measures envisaged in order to avoid, reduce or remedy those effects; and

   (e) a summary in non-technical language of the information specified above.

3. An Environmental Statement may include, by way of explanation or amplification of any specified information, further information on any of the following matters:
   (a) the physical characteristics of the proposed development, and the land-use requirements during the construction and operational phases;

   (b) the main characteristics of the production processes proposed, including the nature and quantity of the materials to be used;

   (c) the estimated type and quantity of expected residues and emissions (including pollutants of water, air or soil, noise, vibration, light, heat and radiation) resulting from the proposed development when in operation;

   (d) (in outline) the main alternatives (if any) studied by the applicant, appellant or authority and an indication of the main reasons for choosing the development proposed, taking into account the environmental effects;

   (e) the likely significant direct and indirect effects on the environment of the development proposed which may result from:
   (i) the use of natural resources;
   (ii) the emission of pollutants, the creation of nuisances, and the elimination of waste;

   (f) the forecasting methods used to assess any effects on the environment about which information is given under subparagraph (e); and

   (g) any difficulties, such as technical deficiencies or lack of know-how, encountered in compiling any specified information.

   In paragraph (e), 'effects' includes secondary, cumulative, short-, medium- and long-term, permanent, temporary, positive and negative effects.

4. Where further information is included in an Environmental Statement pursuant to paragraph 3, a non-technical summary of that information shall also be provided.

Source: HMSO, 1989

2.3 INITIAL CONSIDERATION OF DEVELOPMENT OPTIONS

Depending on the nature of the scheme, the initial planning stages may involve a review of a wide range of proposals in terms both of location and of the precise nature of the development or processes involved. Chapter 7 discusses this further in the context of Integrated Pollution Control (IPC). The outcome of such reviews will usually depend on a range of financial, socio-economic, engineering and other factors. However, it is equally important that the potential environmental impacts of the options are also considered as part of the planning and feasibility stages.

Engineering feasibility studies acknowledge that site topography, ground conditions, access etc. will affect the configuration, engineering design and cost of the scheme. However, it must also be recognised that the environmental characteristics of the locality may necessitate fundamental differences in approach in terms of scheme design, construction and operation.

Acknowledgement of the possible environmental impacts of a planned development at the pre-feasibility stage and the incorporation of environmental criteria within the site selection process comprise an important first stage in the development of a project. Early consideration of these effects before any commitment is made in respect of either the acquisition of the development site or the preliminary scheme design can avoid costly and unexpected delays and costs.

Even the smallest developments might create tangible impacts that can lead to inconvenience, disturbance and annoyance, particularly during the construction phase. The cumulative effect of several small developments in an area can often be significant. As the scale of the development increases, the scope of these impacts broadens both in terms of the disturbances to the social and natural features which characterise the site, and in terms of the area over which these impacts will be felt.

For small-scale schemes, it may be relatively easy to identify possible impacts, and avoidance may consist of no more than appropriate communication and management in accordance with established good practice. For example, for a small infill residential development, existing residents should be informed of the nature of the development, the timescale of the construction works, measures to be taken by the contractor to minimise disturbance, and any features of the development that will enhance the area on completion. Provided that the developer ensures that these measures are properly implemented and responds in a timely manner to any unforeseen impacts that might arise during the course of construction, then the development should proceed in a satisfactory manner. There will, however, be cases where even the smallest of development schemes will need to address significant environmental issues and may be the subject of vehement opposition.

The same principles apply to large development schemes, although the scale and complexity of issues are understandably greater and, in some cases, more difficult to identify since a range of individuals and statutory, voluntary and other organisations will have an interest in the site and its environs.

The location selected may have a significant influence on the construction and operation of the scheme, and the planning stage should allow flexibility in the way the scheme may be adapted to minimise any impacts and take advantage of any enhancement opportunities afforded by the site.

2.4 PRELIMINARY ASSESSMENT OF THE POSSIBLE ENVIRONMENTAL IMPACTS OF THE SCHEME

2.4.1 Preliminary scoping exercise

The environmental assessment should focus on the significant impacts both on and off site that are likely to arise as a result of the development. This requires consideration of the construction and operational phases of the development, together with any post-operational or decommissioning implications, and an understanding of the consequences of these activities on the locality.

Such a preliminary assessment is usually undertaken in advance of widespread consultation, and often during a period when, for commercial or other reasons, the development is not being publicised. Such an exercise will help the developer to anticipate some of the more obvious impacts that may arise and the mitigation measures that may be necessary. This will help to assess whether the scheme is feasible and identify the main interest groups.

A description of the scheme, and an appreciation of the issues involved, will also help to get consultation off to a constructive start.

The importance of the scoping stage cannot be overemphasised. A study of the first 100 EAs undertaken in the UK demonstrated that the poor quality of many Environmental Statements could be traced back to an inadequate scoping exercise, with the result that significant environmental characteristics were not identified and considered in adequate detail during the data-collection stage (Wood and Jones, 1991). This also underlines the importance of putting together an investigatory team with an appropriate breadth of skills. This is also covered in later chapters (e.g. Section 4.3.1).

Box 2.3 Case study: Scoping

Scoping case study

A scoping study for a waste water treatment plant at Fleetwood identified a range of potential impacts, including air and water pollution, noise, odours, traffic and other problems. While many of these impacts could be anticipated, particularly given the proximity of the development to residential areas, an ecological survey of the site and its environs also identified that the loss of reed and willow in the area to the south-east of the works would disturb sites used by birds.

Courtesy North West Water

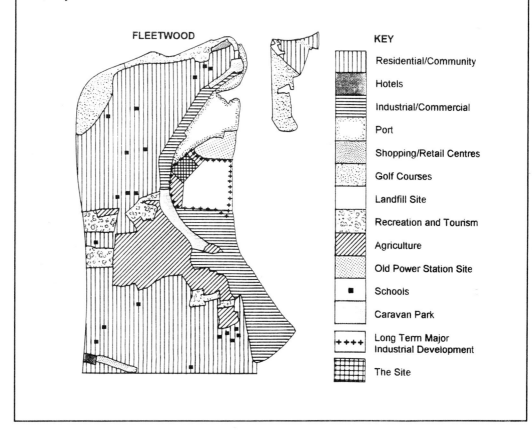

Box 2.4 Project life-cycle

The figure shown below is a simplified representation of the life-cycle of most construction schemes. The main components of the figure relate to:

- the extraction of raw materials
- the use of energy
- manufacture and assembly of products
- the building or construction phase
- the occupational or operational phase
- refurbishment
- demolition
- waste disposal.

These components are highly interrelated: for instance, each will involve the consumption of resources and energy, and each will produce waste. Furthermore, each will have its own transport and labour requirements and secondary effects arising therefrom. The EA must therefore consider the potential impacts of all these components, which form the overall life-cycle of the project.

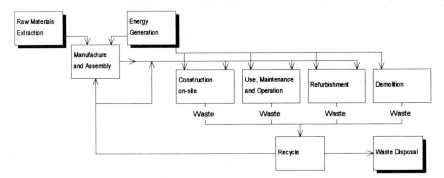

Raw materials
What are the main requirements of the scheme in terms of raw materials?
Where will they be derived from?
Will this activity create new or intensify existing impacts?
How will materials be transported to site?
Does this transport requirement create nuisance, danger or other risks?
Will material winning require special reinstatement?

Energy requirements
What services are required?
Will this be provided by on-site generation?
Will there be on-site fuel requirements, and how will the fuel be stored?
How will fuel reach the site and what are the transport and safety implications?
What permanent services will be required by the operational project?

Manufacture and assembly
Will the project involve extensive import of structural components?
Will an area be required for prefabrication?
Does the project depend on local manufacturing output?

Construction on site
What are the particular characteristics of the site?
How will these influence construction methods?
What operations are involved?
What are the plant, labour and waste implications (housing, accommodation, services etc.)?

Operation
What are the principal resource requirements of the project in its operational phase?
What wastes will arise during operation?
What are the transport implications of this?
Who will the project be used by?
How and when will they require access to the site?
What operations take place on the site?
What emissions will arise?

Maintenance and refurbishment
What operations are required to maintain the project in an operational state?
Will this require any special materials, plant or labour?
How often will this be required?

Demolition
Does the project have an expected finite life?
Will it require dismantling or demolition?
How will this be done?

Waste disposal
What wastes arise from the project during routine operation?
What wastes could arise during periodic maintenance and refurbishment?

2.4.2 Preliminary consultation and data collection

The purpose of this stage of consultation and data collection is to identify the environmental interests at the site and surrounding areas, and establish the availability of relevant data and information. Consultation is covered in greater detail in Section 2.5.

The types of issues that arise will relate both to the site and the to nature of the proposed development. For the latter, an understanding of the activities involved throughout the lifecycle of the project will help to focus the scoping exercise and assist in the consultation process. Box 2.4 illustrates some of the questions that will need to be considered early in the planning stages of the project.

The potential construction impacts are clearly influenced by the overall scheme design, and the building and engineering methods employed in its construction. Chapters 3–9 of this document provide a detailed description of the construction activities involved in different types of project and in a variety of environments. The project may involve a range of different activities, some of which are listed in Box 2.5.

Box 2.5 Activities associated with construction projects

Site investigation	Trial holes, soil and rock sampling, test wells, ground bearing and other physical tests etc.
Demolition and site clearance	Removal of existing buildings and structures, temporary access roads, removal of topsoil, clearance of trees and shrubs, removal of existing services
Geotechnical and other specialised processes	Various methods by which the properties of the ground are modified, including grouting, drainage, diaphragm walls and other impermeable barriers, remediation of contaminated land
Tunnelling	Boring, blasting and spoil disposal
Earthworks	Dredging, excavation for foundations and basements etc., re-profiling site contours etc.
Piling	Variants include bored or driven; pre-formed or cast in situ
Superstructure	Fabrication of pre-formed components, casting of structures in situ, laying of brickwork, blockwork, stone, rip-rap etc.
Services	Drains, water and gas supplies etc.
Roads and paving	Access roads including public highways, loading areas, car parks
Fitting out	Installation of plant, internal services and cladding, testing etc.
Landscaping	Planting, seeding, replacement of top soil etc.
Testing and commissioning	Controls, manuals, monitoring to optimise operation etc.

A fuller description of these activities is beyond the scope of this document, and they are covered extensively elsewhere (Powell, 1979; Blake, 1989; CIRIA, 1993b). The implications of many of these activities are also discussed in general terms in Chapter 3.

The assessment of potential impacts can only be achieved through an understanding of the operations employed and their associated risks. This further underlines the need to integrate the activities of those responsible for the design and those responsible for the EA.

An understanding of the activities on and off site will also enable opportunities for environmental enhancement to be identified (see also Section 2.6).

2.4.3 Types of impact

It is difficult to categorise precisely the possible impacts of a development, since these will largely depend on the characteristics of the locality. The EA Regulations require impacts to be identified in terms of their effects on (HMSO, 1989):

- human beings
- flora
- fauna
- soil
- water
- air
- climate
- the landscape
- the interaction between any of the foregoing
- material assets
- the cultural heritage.

These have been simplified into three categories for the purpose of this document. However, owing to the complexity of and interactions between these components, it is acknowledged that other categorisations may be equally appropriate.

Physical/chemical

Physical effects include ground stability problems and long-term geomorphological effects. Examples of these effects and their causes include: flooding of valleys affecting slope stability; river improvement works affecting flow velocity, erosion and sediment transport; and changes to groundwater movements through deep excavations, dewatering or the introduction of extensive lenses of impermeable material (e.g. a landfill lining).

Chemical effects are primarily those involving discharges to land, water or air. However, it is acknowledged that many of the criteria for setting discharge standards for these materials relate to human health, aquatic life, uptake of ground pollutants in crops etc.

These effects can be acute, chronic, localised, regional or global as indicated in Table 2.1. Their effects can therefore manifest themselves in humans, plants, and animals.

Table 2.1 Examples of physical and chemical effects

Effect	Examples
Acute	Fish kills from toxic spills
Chronic	Bronchial disorders from nearby stack emissions Food chain effects Asbestosis
Localised	Vehicle emissions
Regional	SO_2 contribution to acid rain Dioxin emissions
Global	CO_2 contribution to global warming

Human effects

Human effects are defined in this document as covering all the impacts on human beings, other than chemical effects such as those described immediately above. These include disturbances such as noise, odours, inconvenience such as traffic congestion, and severance effects caused by roads. Also included are social and cultural impacts such as those on landscape, and features of geological and historic interest.

Nature conservation

Many of the physical and chemical effects of a scheme will have implications on plant and animal life. For instance, changes to groundwater patterns may lead to the loss or proliferation of a particular plant species. Water pollution may have chronic or acute effects on aquatic flora and fauna, while air pollution can have marked effects on plant life, sometimes leading to irreparable damage to established vegetation. Other disturbances may disrupt breeding or migratory patterns, while the development itself may obstruct important wildlife corridors. Many of these effects may be exacerbated or, conversely, minimised according to the timing of construction and operational activities. Further guidance on nature conservation issues is provided in a report by English Nature (1994).

Construction activities can also damage features of geological and geomorphological value. For example, land forms or the physical processes maintaining them can be damaged by construction activities, and important rock exposures can be covered or destroyed. However, other activities can create valuable exposures or lead to the enhancement of existing sites.

2.4.4 Special sites

The general principles involved in establishing the environmental characteristics of sites can be applied to all locations. However, in certain instances, there may be additional important factors to be considered by virtue of the special characteristics of a site.

Perhaps the best-known examples are the Sites of Special Scientific Interest (SSSIs) where valuable natural characteristics are identified. Further examples of special designations are listed in Box 2.6.

The definition of Aquifer Protection Zones by the NRA (1992) and associated planning constraints, and the proposals for a register of contaminative land uses (DoE, 1992b) illustrate other aspects of the site that may have significant influence on the design.

Box 2.6 Nature conservation and additional designations

Context	Designation
International sites	Special Protection Areas Ramsar Sites World Heritage Sites* Bisophere Reserves Special Areas of Conservation
National sites	National Nature Reserves Sites of Special Scientific Interest Nature Conservation Review Sites (Grade 1)* Marine Nature Reserves
Other sites	Ancient woodland* Local Nature Reserves County trust reserves* Local authority reserves* Local authority designated sites* County trust designated sites* Other non-government organisation reserves* (for example Woodland Trust, Royal Society for the Protection of Birds etc.)

*Non-statutory sites

Examples of additional designations and descriptions that should be considered

National Parks
Areas of Outstanding Natural Beauty
Heritage Coast
Local landscape designations*
National Water Council (NWC) classification
EC Salmonoid and Cyprinid Fisheries
Aquifer Protection Zones*
Environmentally Sensitive Areas (ESAs)
Historic Parks and Gardens*
Tree Preservation Orders
Conservation Areas
Recreation Areas (for example Country Park)

*Non-statutory sites
(Note that a site can be the subject of more than one designation)
from English Nature (1993)

2.5 ESTABLISHMENT OF THE ENVIRONMENTAL CHARACTERISTICS OF THE SITE

2.5.1 General requirements

A combination of site visits, literature reviews, gathering of available information and consultation with interested parties should provide an indication of the type of issues that are likely to be important on a scheme-specific basis. Some impacts will be obvious, and the success of the scheme will depend on appropriate mitigation measures. However, there will usually be a further category of impacts whose significance will not be so clear at the outset. These will require additional research in order to establish their significance and identify any necessary mitigation measures that need to be taken. Depending on the nature of the issues involved, the requirements of additional specialist skills will also be identified to ensure that the

EA/design team has the necessary expertise to deal with all the issues identified in the scoping exercise.

The preliminary assessment of the possible impacts of the project will help to identify, in general terms, some of the impacts that may be expected to arise. However, a detailed assessment of the environmental characteristics of the site will require:

- bringing together appropriate engineering and environmental skills into the planning and design team at an early stage
- further consultation with interested parties likely to be affected by the development
- ensuring productive interaction between all parties involved
- further scoping and identification of possible environmental issues
- gathering of baseline data for the site
- iterative modification of the scheme proposals to mitigate any impacts and, where appropriate, provide environmental enhancement.

Some of the information required for a particular project is likely to exist already: in published reports, in archived files, or with individuals. Within reason, every effort should be made to identify the existence of such information. While not complete, nor in an ideal format, it may yield valuable background information about the site. Making maximum use of such data will also reduce the overall cost to the developer.

2.5.2 Consultation

Discussions with consultees will often permit access to expertise and knowledge of local conditions that might otherwise not be possible. Particular data-collection requirements, including any necessary timing requirements or specialist skills, will also be identified together with possible mitigation measures.

For many schemes, such consultation is statutory (see HMSO, 1989). Early consultation is welcomed by consultees, committees and local interest groups and will help to ensure that the developer is fully aware of the basis of concerns and can take measures to avoid anticipated impacts wherever appropriate and practicable. Statutory consultees are required to make information available to assist developers in the preparation of an ES (Regulation 8 of the Town and Country Planning (Assessment of Environmental Effects) Regulations 1988). Consultation by the developer's competent authority on the application for development consent is a statutory requirement.

For large-scale schemes, consultations associated with the EA may themselves be a major undertaking, requiring commitment and resources from those undertaking the assessment, the scheme promoters, statutory consultees and the public. To be effective, the consultations must include a willingness to seek advice from interested parties, to pay attention to what they say, and to act upon it. If undertaken properly, the result will be a better-quality assessment and a smoother implementation of the proposals. If mishandled or entered into half-heartedly, the consultation exercise will only serve to increase suspicion and could lead to a substantial delay in achieving the project objectives.

Consultation should continue throughout the data-collection stage and beyond. For large or sensitive schemes, consultation groups should be maintained to discuss issues of concern and the success of mitigation measures.

A list of consultees and their principal areas of interest is shown in Table 2.2. It must be stressed that the table is not exhaustive. Furthermore, whereas national bodies are easily identifiable, it is impossible to list all the local and community interest groups that might have an interest in particular projects. Further information is provided in publications such as *Who's Who in the Environment* (Environment Council, 1992) or in various environmental directories (e.g. Frisch, 1992).

Table 2.2 Possible consultees

Areas of principal interest	Consultee	Areas of principal interest	Consultee
Traffic/ transport	Department of Transport Secretary of State for Wales Local authority British Rail	Pollution	Local authority (planning and environmental health) HMIP NRA
Heritage	English Heritage CADW (Wales) National Trust Historic Buildings and Monuments Commission Landmark Trust Local authority (archaeologists and conservation officers) Royal Fine Arts Commission County or local museums	Nature conservation	English Nature Countryside Council for Wales Nature Conservancy Council for Scotland Royal Society for the Protection of Birds County wildlife trusts Council for the Protection of Rural England Local authority Local conservation organisations Royal Society for Nature Conservation Wildfowl and Wetlands Trust Woodland Trust British Association for Shooting and Conservation Marine Conservation Society National Trust British Trust for Ornithology Forestry Commission Botanical Society of the British Isles Scottish Wildlife Trust National Parks National Rivers Authority
Existing buildings	Local authority Development associations Civic Trust Chamber of commerce Residents association	Flooding and drainage	National Rivers Authority Internal drainage boards Local authority
Utilities	Local water and sewage companies Local electricity company Telecommunication company Electricity generation companies Local authority British Gas National Grid British Pipelines Agency Ministry of Defence Local authority Cable TV companies	Amenity and leisure	Sporting associations Youth groups Tourist boards Local authority Sports Council CPRE Ramblers Association Royal Yachting Association The Sports Council Long Distance Walkers Association Cyclists Touring Club Theatres Trust
Climate	Meteorological Office MAFF Local authority	Planning issues	Local planning authority MAFF NRA
Fisheries	Ministry of Agriculture, Fisheries and Food National Rivers Authority Sea Fish Industry Authority Department of Agriculture and Fisheries for Scotland River Purification Boards Riparian owners Sea Fisheries Committee Local angling associations Fisheries Statistical Unit (Scotland) The Scottish Salmon Fisheries Inspector	Coastal issues	National Rivers Authority Local authority Ministry of Agriculture, Fisheries and Food English Nature Countryside Council for Wales Countryside Commission Private owners Marine Information Advisory Service Meterological Office Local port authorities Department of Transport Nature conservation groups Council for the Protection of Rural England
Water quality	Water companies National Rivers Authority Waste regulatory authorities Local authority	Sediment quality	Ministry of Agriculture Fisheries and Food Waste regulation authority

continued/...

Table 2.2 (continued)

Areas of principal interest	Consultee	Areas of principal interest	Consultee
Soil	Ministry of Agriculture Fisheries and Food Soil Survey of England and Wales Welsh Office Scottish Office	Geological interest	English Nature Countryside Council for Wales British Geological Survey
		Hydrology/ geomorphology	National Rivers Authority Internal drainage boards English Nature Countryside Council for Wales British Waterways Nature conservation groups
Air quality	Her Majesty's Inspectorate of Pollution Local authority (environmental health officers)	Human issues	Parish council Local authority Department of Employment Chamber of commerce Local planning authority
Health and safety/risk	Health and Safety Executive Local authority (environmental health officer) Waste disposal and regulatory authorities	Emergency services	Emergency services
Land use	MAFF Local authority National Farmers Union Country Landowners Association Crown Estates Commissioners CPRE National Parks Authority British Coal Corporation Waste disposal authority Landowners Ordnance Survey British Geological Society	Landscape	Countryside Commission Countryside Council for Wales National Rivers Authority CPRE Local authority National Parks Authority
Tourism	Local authority English Tourist Board Wales Tourist Board Sports Council Regional Tourist Board	Navigation	Port Authority Harbour Commissioners National Rivers Authority British Waterways Trinity House Department of Transport

Note:
1. This table is not comprehensive, particularly in respect of local interests.
2. References to local authorities include county, district, borough, city and metropolitan boroughs. Where specific departments are indicated, this should not be taken to suggest that other departments will not have an interest.
3. English Nature, the Countryside Commission and their equivalents are statutory consultees for all projects.

2.5.3 Baseline data collection

One of the purposes of the scoping exercise is to identify the main issues of concern, and the data collection should concentrate on these issues.

Owing to the potentially vast range of issues involved, the developer must ensure that the data-collection exercise is undertaken by a balanced team containing the necessary skills and specialisms appropriate to the scheme, locality and issues concerned.

Given the resources involved in the data-collection exercise, which may involve extensive field data collection, together with the costs of any associated laboratory or computer analyses, careful attention should be given to the timescale (and timing), parameters, localities, frequency and desired precision of the information collected with particular reference to the way in which it is to be analysed and interpreted.

The environment is a complex arrangement of interrelated systems. Such systems are notoriously difficult to define and model, owing to their inherent variability. Not all environmental characteristics can be measured directly. Many require a subjective approach. Even where measurements are possible, extensive monitoring is required in order to build up an accurate picture in terms of variability, seasonality and trend.

Many specific examples are described in context in Chapters 3−9. Some of these are illustrated in Box 2.7.

Box 2.7 Examples of data subject to seasonal or other variations

---

The water cycle

- river flows
- water table
- rainfall
- water temperature

The first three of these are clearly related. If possible, river flow data should be studied over several years to establish low or high flow conditions. In many cases, this information will already be available from the NRA. A knowledge of low river flows would, for instance, be important in establishing the assimilative capacity of a watercourse to receive waste or where it is to act as a source of water.

Nature conservation

- breeding seasons
- flowering seasons

Owing to the migratory, breeding and feeding patterns of birds and other animals, surveys will need to be planned at appropriate times of the year in order to establish an accurate picture of the types and extent of various species that are native to the site and its environs. In many cases, the sampling periods for different species vary, and the most intensive investigations will be of little or no value if they are incorrectly timed. Further discussion of this important area is provided in Chapter 7 and in the report by English Nature (1994).

Noise

- background noise
- wind noise

Background noise can be expected to vary significantly. Depending on the location, there may be strong daily, weekly or seasonal patterns. Where the background noise arises from natural sources such as wind noise, measurements are more difficult to make. Further discussion is provided in Chapters 5, 7 and 8.

Dominant local features in system

Often, the baseline conditions for a particular characteristic will be dominated by a limited number of features. Rainfall runoff for a sub-catchment may be highly dependent on a particular developed area, whose contribution might be most significant during, for instance, heavy summer storms. Foul water draining to, say, a proposed treatment works might be dominated by a single industrial source where, for instance, balancing tanks are emptied at particular times of the day. In certain industries, discharges may even exhibit marked seasonal variations. The background noise level in an area might be due to a single, albeit major, manufacturing source whose cessation might expose numerous other sources of noise that are usually masked. If possible, sampling should include periods where such manufacturers are closed.

Tourism

Tourism will cause a significant change in seasonal populations in terms of water consumption, waste production, traffic etc. In many localities, a large influx of visitors occurs outside the traditional summer vacation periods. These occasions include major sporting events, festivals and other traditional events.

2.6 IDENTIFICATION OF POSSIBLE IMPACTS AND APPROPRIATE MITIGATION MEASURES

2.6.1 Identification and prediction of impacts

Definition of the project activities, analysis of the baseline data conditions and consultation will help to assess the likelihood, nature and extent of the impacts associated with the proposed scheme, and hence indicate the range of measures required to avoid, minimise or compensate for these impacts. It is likely that a range of issues will be identified which could:

- affect the natural, human, chemical and physical environments
- take place within the confines of the site and beyond
- occur during the construction, operational and post-operational phases
- be reversible or irreversible
- be positive or negative
- cause primary and secondary impacts.

Prediction techniques are covered in later chapters.

Many of these impacts will be commonplace, and avoidance will be a matter of good practice. However, some will require specialised mitigation measures.

In addition to identifying potential negative impacts, the initial phases of the assessment will also identify particular features of the site which may provide conservation and enhancement opportunities. The integration of such features into the design will often yield commercial benefits.

2.6.2 Mitigation measures

Mitigation will consist of a number of related actions, many of which would consist of no more than ensuring effective management and control of site or operational activities. At the other extreme, some measures will form a permanent and integral element of the development and its operation.

In addition to specifying mitigation measures, appropriate means of monitoring their effectiveness should be considered. Despite the best intentions, a lack of effective implementation during construction will lead to unnecessary conflict and may lead to delays in the scheme. Furthermore, commitment to mitigation should also include post-project monitoring of the performance of any permanent mitigation measure.

The suitability and acceptability of proposed measures should be discussed with consultees. Compensatory measures or substitution of natural habitats might not always be acceptable. Also, certain measures might themselves create secondary effects or interact with other elements in the locality in an unexpected way.

Data collection, consultation and other investigations might lead to a full understanding of the potential impacts of a scheme at its various stages. If the impact of the scheme is to be kept to a practical minimum, active measures will need to be taken to address the various issues raised. This introduction has demonstrated that impacts can occur at any or all of the stages in the life-cycle of a project. Chapters 3–9 of this document illustrate this further.

Preventative or mitigation measures can take many forms. Different types of measures are listed below.

- Select alternative location
- Modify working practices during construction
- Development designed to minimise operational impacts
- Effects of impacts from site minimised by off-site measures.

Box 2.8 Case study: Mitigation

An Environmental Assessment undertaken for a 411 kilometre-long pipeline between Grangemouth, in the Central Region of Scotland, and Stanlow, in Cheshire, identified many ecologically sensitive sites which could not be avoided owing to other constraints. As a consequence, a high priority was given to the care of the environment, and mitigation measures were implemented at sensitive sites.

One such site was a raised bog at Harnhead, near Avonbridge in Scotland. At the site it was important to ensure that the surface of the bog, dominated by mosses, heather and hairstail cotton-grass, was carefully reinstated. This was done by narrowing the working area and by confining all vehicles to a temporary road made of wood. In addition, the vegetation was removed from the width of the pipe trench only. It was removed as large turves, and the turves were wrapped during storage to prevent drying.

An inspection of the site approximately six months after the work had been completed showed that the replaced turves were growing well. Two years later it is very difficult to see the line of the pipe.

Photos courtesy RSK Environment

Initial condition

Six months later

General examples of each of these approaches are provided in Box 2.9. Further details are provided in Chapters 3–9. Arrangements should also be made to monitor and review the predictions made in the Environmental Statement in terms of their accuracy and the success of specific mitigation measures.

Box 2.9 Examples of mitigation measures

---

1. SELECT ALTERNATIVE LOCATION

Transfer of the proposed development to another location might not always be an option. For property development, or the location of a new manufacturing base, it may be a possibility. However, activities such as minerals extraction are clearly influenced by the natural occurrence of such materials, while replacement of flood or sea defence structures is a site-specific decision.

There may be some scope for adjusting the layout of the site so as to minimise damage, risk or nuisance. Examples might be avoidance and protection of sensitive natural systems, landscaping and alignment to minimise visual and noise impacts.

2. MODIFY WORKING PRACTICES DURING CONSTRUCTION

Many impacts can be minimised by fairly simple measures, particularly where the impact is one of disturbance or nuisance. Pollution can also be avoided by taking appropriate measures. Examples of this include:

- limiting hours of working
- utilising quiet plant
- avoiding audible vehicle reversing alarms in favour of other methods
- ensuring that vehicle exhausts do not point down if dusty conditions predominate
- providing vehicle wheel washers
- protecting site against vandals
- careful storage of fuel oils and chemicals
- special transport arrangements

(see also off-site measures).

More fundamental changes might include the specification of bored as opposed to driven piles. Also, where the construction requires the use of explosives, blasts can be configured to create minimum disturbance and vibration by such measures as sequenced detonation. Further details are provided in Chapters 5 and 8.

3. CONTROL IMPACTS ARISING FROM DEVELOPMENT

The operational long-term effects of the development can be minimised in several ways.

Noise and vibration

Where the development includes operational plant, noisy items such as generators, compressors and other plant can be housed in sound-insulated buildings. Landscaped bunds could also be provided to minimise the projection of noise.

Pollution and waste

The reduction of impacts arising from discharges to land, water and air broadly follows three themes: containment; collection, treatment and disposal; dilution and dispersal.

The impermeable layers required on landfill sites are an example of the first of these types of approach, although most modern methods will also include means of dealing with the leachates and gases produced during this process. Facilities for the containment of accidental spills might also be required in certain circumstances.

Collection and treatment methods are widely employed in a range of developments. The collection of polluted surface water from vehicle hardstandings and the provision of oil traps is one example. Treatment of waste will usually involve the disposal of various residues.

Dilution and dispersal employs the assimilative and dispersal properties of the receiving medium. Examples of this approach include: the discharge of treated sewage effluent within controlled consents; the raising of chimney stacks to ensure that gaseous emissions are adequately dispersed by wind; and the spreading of sewage sludge onto land at prescribed rates in accordance with crop uptake rates.

Further discussion is provided in Chapter 7, which outlines the concept of Integrated Pollution Control.

continued/...

Box 2.9 (continued)

Visual aspects

Architectural and visual standards are a matter of taste and subjectivity (see Chapter 3). However, the visual impact of the development can often be minimised by well-planned landscaping, the use of local materials and styles, and attention to form, colour and texture. The nature of many developments, particularly those enclosing processes such as power stations or incinerators, may necessitate large monolithic structures. In this case, the emphasis will probably be on minimising the visibility of the structure by landscaping and screening.

4. OFF-SITE MEASURES OF CONTROL

In circumstances where it is not possible to restrict the scope of any impact of the development, mitigation measures can be employed off site to reduce the severity of the impact.

For instance, the planting of trees can offer many benefits (Coppin and Richards, 1990):

- a recreational feature
- a visual feature or screen
- a natural habitat
- wind shelter to offset any wind effects caused by the development
- reduction of noise
- consumption of CO_2 e.g. to help offset that produced by fossil fuel combustion.

However, careful selection of species is required. Also, such measures should not be viewed as necessarily providing compensation for the destruction of an existing habitat.

Off-site solutions are sometimes acceptable where valuable habitats are threatened by a development. With suitable advice, habitat translocation or recreation may be possible.

Traffic impacts off site can be minimised by providing adequate signposting and avoidance of narrow congested roads by delivery vehicles etc. Also, though a more costly measure, serious anticipated traffic-related problems may be offset by local road improvement schemes.

Overhead power transmission lines are highly visible. However, where they present the only viable option, measures can be taken to avoid other impacts such as bird strikes (see Chapter 7).

2.6.3 Enhancement opportunities

Environmental enhancement can often be achieved with little or no additional cost. The incorporation of natural features in the final design, the creation of habitats, the provision of public facilities such as linear parks, footpaths, wildlife observation facilities and, for significant projects, visitor centres can provide opportunities to realise benefits that go beyond those associated with the project itself.

Often, these opportunities may arise during the construction phase. Construction operations such as excavations for clay or aggregates can provide opportunities for ponds or lakes, which may provide new wildlife habitats and provide features that can enhance the development in other ways. The additional effort required may be small, particularly as reinstatement of the site may otherwise be required (see Box 2.10).

Box 2.10 Case study: Enhancement

Several large reservoirs have been constructed in the UK whose primary function is flood alleviation. However, such facilities are increasingly being viewed as providing opportunities for nature conservation, amenity, water quality enhancement and recreation. Shoreline and aquatic planting can provide shelter and breeding grounds for birds. Islands can provide small nature reserves, and lake depths can be adjusted to be conducive for fish breeding and feeding requirements. Recreation facilities, whether in the form of walkways or launching and other facilities for sailing, can also be provided. Wilen Lake in Milton Keynes is a good example of a reservoir with many of these features.

Photo courtesy of Milton Keynes Development Corporation

2.6.4 Contingency plans

The construction of projects is subject to numerous external influences. Weather effects are probably the most significant of these. The construction plan for the scheme should therefore include contingencies that are operated in times of flood. Accidents and vandalism on and off site should be anticipated, particularly those involving polluting or hazardous materials. Accidents and emergencies range from damage to existing services, or spillages of chemicals, to the failure of a dam. Where the scale of operations or the potential severity of effects justifies such an approach, these plans should include the emergency and pollution control services.

The operation of completed developments might also include processes or activities that pose risks, particularly where these involve transport and storage of polluting or hazardous materials. These should be identified and appropriate contingency plans prepared.

2.7 REFERENCES

Blake, L.S. (1989)
Civil Engineers Reference Book
Butterworth

BSI (1992)
Environmental Management Systems
British Standards Institute, Milton Keynes

CIRIA (1993b)
Environmental handbook for building and civil engineering
SP98 - Construction Phase

Coppin, N.J. and Richards, I.G. (1990)
Use of Vegetation in Civil Engineering
CIRIA/Butterworths, Book 10

DoE (1992a)
Planning controls over Sites of Special Scientific Interest
Circular 1/92

DoE (1992b)
Register of contaminative uses – Consultation Paper
Department of the Environment, London

DoE/Welsh Office (1988)
Environmental Assessment
Circular (15/88) / (23/88)

English Nature (1994)
Nature conservation in Environmental Assessment
Final Report
English Nature (Peterborough)

Environment Council (1992)
Who's Who in the Environment
Environment Council (London)

Frisch, M. (1992)
Directory of the Environment
Green Print (London)

HMSO (1988)
Town and Country (Assessment of Environmental Effects) Regulations 1988

HMSO (1989)
Environmental Assessment, A guide to the procedures
HMSO

NRA (1992)
Policy and practice for the protection of groundwater
National Rivers Authority (Bristol)

Powell, M.J.V. (1979)
House Builders Reference Book
Butterworth

Wood, C. and Jones, G. (1991)
Monitoring environmental assessment and planning
EIA Centre, Department of Planning & Landscape, University of Manchester

CIRIA Research Project 424

Environmental Assessment

A guide to the identification and mitigation of environmental issues in construction schemes

3. Built development

Michael Renger, Partner, Nabarro Nathanson

3 Built development

3.1 TYPE OF DEVELOPMENT CONSIDERED

Most design and construction companies will become involved with some form of building development as part of their activities. Indeed, for many, this comprises the bulk if not the whole of their workload.

Building development is defined in this chapter as encompassing a range of schemes, from simple house extensions to extensive mixed developments including residential developments, office and retail schemes, leisure complexes and light industrial development. In the public's mind, the construction industry is probably typified by general building development of this nature, since it is to such development that they are most often exposed. The construction of such structures has much in common with other types of development. Once they are built, their impacts arise from the use made of the buildings — whether domestic, commercial, retail or leisure — together with their associated activities — the consumption of resources, the demand on energy, transport, waste and other services. This chapter does not consider the operation of buildings or structures associated with manufacturing or processing or other uses which, themselves, involve specific pollution or special waste problems. However, much of the chapter will be relevant to the construction of such facilities. Also, the chapter does not address the assessment of the major infrastructure works such as roads and pipelines. These are discussed in later chapters.

3.2 THE LEGAL FRAMEWORK

As mentioned in Chapter 2, European Community Directive No 85/337 lists two types of project, one for which an Environmental Assessment (EA) must always be carried out (Annex I) and the other for which an EA may be required where the project is likely to have significant environmental effects (Annex II). There are no categories of built development within the scope of this chapter that are included in Annex I, and for which an EA must always be carried out. The Annex II operations, however, include three such categories of development:

- an industrial estate development project
- an urban development project
- a holiday village or hotel complex.

The European Directive was transposed into English law by the Town and Country Planning (Assessment of Environmental Effects) Regulations 1988 ('the Regulations'), with Schedules 1 and 2 reiterating the categories of development contained in Annexes I and II. Schedule 2 reiterates the above three categories of development, and the Joint Circular from the Department of Environment (Circular 15/88) and Welsh Office (Circular 23/88) contains the respective Secretary of State's guidance as to when an environmental assessment will be required for Schedule 2 projects. The primary source documents are highlighted in Box 3.1.

Box 3.1 Source documents

> European Community Directive 85/337/EEC – The Assessment of the Effects of Certain Public and Private Projects on the Environment.
>
> Town and Country Planning (Assessment of Environmental Effects) Regulations 1988 (SI 1988/1199).
>
> The Department of the Environment Circular 15/88 (Welsh Office Circular 23/88).
> 'Blue Book' *Environmental Assessment: A guide to the procedures* issued by the Department of the Environment and Welsh Office.
>
> Town and Country Planning (Assessment of Environmental Effects) (Amendment) Regulations 1990 (SI 1990/367).
>
> Town and Country Planning (Assessment of Environmental Effects) (Amendment) Regulations 1992 (SI 1992/1494).
>
> Department of the Environment Consultation Papers (e.g. 1992 Consultation Paper on Environmental Assessment and Planning).

The need for EA for Schedule 2 projects will be determined case by case but Government guidance indicates that, in general terms, EA will be required for Schedule 2 projects in three principal types of case:

- major projects that are of more than local importance

- occasionally, projects on a smaller scale, which are proposed for particularly sensitive or vulnerable locations

- in a small number of cases, projects with unusually complex and potentially adverse environmental effects, where expert and detailed analysis of those effects would be desirable, and would be relevant to the issue of principle as to whether or not the development should be permitted.

The Secretary of State's view is that the number of projects falling within these categories will be a small proportion of all Schedule 2 projects. The basic test to be applied in determining the need for a formal EA in a particular case is whether there is a likelihood of significant environmental effects, and not the amount of opposition or controversy to which a project gives rise, except to the extent that the substance of objectors' arguments indicates that there may be significant environmental issues.

In the Appendix to the Circular, the Secretary of State provides indicative criteria and thresholds to assist local planning authorities in deciding whether a Schedule 2 project should be subject to EA. The relevant extracts from the Appendix for built development are shown in Box 3.2.

Section 15 of the Planning and Compensation Act 1991 empowers the Secretary of State to extend those categories of development for which a formal EA may be required.

As indicated earlier, inclusion of the category of development within Schedule 2 does not make the preparation of an EA mandatory, but rather gives discretion to the relevant authority as to whether an EA should be prepared. In the UK the relevant authority is the local planning authority or the Secretary of State. The Government has issued guidance on the operation of this system, not only in the Department of Environment Circular 15/88 referred to previously, but also in the form of a non-statutory booklet entitled *Environmental Assessment: A guide to the procedures* (HMSO, 1989).

Primary legislation of direct interest to developers of built developments is listed in Table 3.1.

Box 3.2 Advisory thresholds for EA

Industrial estate development projects

Industrial estate developments may require environmental assessment where:

(i) the site area of the estate is in excess of 20 hectares; or

(ii) there are significant numbers of dwellings in close proximity to the site of the proposed estate: e.g. more than 1,000 dwellings within 200 metres of the site boundaries.

Smaller estates might exceptionally require environmental assessment in sensitive urban or rural areas, particularly if associated with other works (e.g. roads, canalization projects, flood relief works) which are listed in Schedule 2 (to the Regulations).

Assessment of an industrial estate proposed as an infrastructure project will not necessarily remove the need for assessment of individual industrial installations to be provided within the estate. These might require environmental assessment, if they fall within Schedule 2 (to the Regulations) and are likely to give rise to significant environmental effects which need to be appraised separately from the effects of the estate as a whole.

Urban development projects

Redevelopment of previously developed land is unlikely to require environmental assessment unless the proposed use is one of the specific types of development listed in Schedules 1 or 2 (of the Regulations) or the project is on a very much greater scale than the previous use of the land.

The need for environmental assessment for new urban development schemes on sites which have not been intensively developed should be considered in the light of the sensitivity of the particular location. Such schemes (other than purely housing schemes) may require environmental assessment where:

(i) the site area of the scheme is more than 5 hectares in an urbanised area; or

(ii) there are significant numbers of dwellings in close proximity to the site of the proposed development, e.g. more than 700 dwellings within 200 metres of the site boundaries; or

(iii) the development would provide a total of more than 10,000 square metres (gross) of shops, offices or other commercial uses.

Proposals for high-rise development (e.g. over 50 metres) are not likely to be candidates for environmental assessment for that reason alone; but this may be an additional consideration where one or more of the above criteria is met.

Smaller urban development schemes may require environmental assessment in particularly sensitive areas: e.g. central area redevelopment schemes in historic town centres.

The need for environmental assessment in respect of proposals for major out-of-town shopping schemes should also be considered in the light of the sensitivity of the particular location. For such schemes, a floor area threshold of about 20,000 square metres (gross) may provide an indication of significance.

Other infrastructure projects (e.g. a holiday village or hotel complex)

No specific thresholds or criteria are indicated for holiday villages or hotel complexes. However, a broad indication of likely environmental effects may be given by the land requirement for an infrastructure project. Projects requiring sites in excess of 100 hectares may well be candidates for environmental assessment.

Regulation 5 of the 1988 Regulations enables the developer to request the local planning authority to state in writing whether they are of the opinion that the proposed development would require an EA. The request can be made in the form of a letter but should be accompanied by a plan sufficient to identify the land, a brief description of the nature and purpose of the proposed development and other possible effects on the environment and such other information or representations as the developer would wish to make in relation to the application.

Table 3.1 Other statutory sources

Date	Provision	Purpose/relevance
1949	National Parks and Access to the Countryside Act	Access to open countryside, public rights of way, National Parks, nature conservation
1974	Control of Pollution Act	Noise control
1979	Ancient Monuments and Archaeological Areas Act	Schedule of ancient monuments
1980	Highways Act	Public rights of way, new road schemes
1981	Wildlife and Countryside Act	Nature conservation, SSSIs
1982	Derelict Land Act	Derelict land grants, land clearance areas
1985	Wildlife and Countryside (Amendment) Act	SSSI notification
1990	Environmental Protection Act	Establishment of English Nature and Countryside Council for Wales
1990	Planning (Hazardous Substances) Act	Storage of hazardous substances
1990	Town and Country Planning Act (as amended)	Requiring planning permission
1991	Land Drainage Act	Works affecting drainage
1991	Planning and Compensation Act	Secretary of State's power to extend categories of development for which EA required
1991	Water Industry Act 1991	Powers and duties of privatised water and sewerage undertakers
1991	Water Resources Act	Powers and duties of NRA
1992	Protection of Badgers Act	Protection of badgers
1993	Clean Air Act	Smoke, dust and fumes
1993	Noise and Statutory Nuisances Act	Noise emitted in the street

The local planning authority are obliged to provide their written response within three weeks of the request (or such longer period as may be agreed in writing between the parties) and if the authority's opinion is that an EA is required a written statement should be provided giving clearly and precisely their full reasons for that conclusion. The developer does have a right of appeal to the Secretary of State if he is not satisfied with the opinion expressed by the planning authority. Reasons advanced by planning authorities for requiring an EA of Schedule 2 projects might include:

- projects impinging on important views
- potential interaction with local airport
- loss of high-quality agricultural land
- inadequacy of proposed drainage and its effect on the watertable
- generation of road traffic
- impact on Area of Special Archaeological Significance.

It should be emphasised that the local planning authority, in considering whether or not an EA is appropriate for a Schedule 2 project, should only consider the question of whether the particular project is likely to give rise to significant environmental effects. It should not have regard to, or be persuaded by, the level either of local objections or of support that the project is likely to attract. Local opinion will be a relevant consideration when determining the planning application itself but is not a factor to be taken into account in determining whether or not an EA is required in law.

3.3 VOLUNTARY ENVIRONMENTAL IMPACT ASSESSMENTS

It has already been seen that the categories of built development that are subject to a legal requirement to undertake an EA are few. However, an increasing number of developers are voluntarily undertaking EAs as an integral part of the design and planning of a development project. The question of whether it is appropriate to volunteer an EA is a matter of degree, and no arbitrary rules can be set out. The size of a development, while relevant, is not in itself a determining factor in considering an EA. An EA would often be appropriate for the smallest built development project if that project were to be sited in or adjacent to a Site of Special Scientific Interest, whereas a large-scale urban regeneration project may have no significant impact upon the environment. Every form of built development is capable of being subject to an EA. While a local planning authority can require additional information to be provided in relation to the planning application, it cannot require the submission of an EA unless the project is of a type that falls within Schedule 2 to the Regulations. However, a developer may undertake a full EA to demonstrate to the planning authorities and to the local residents the likely impact of the proposed development upon the environment. Where an ES is submitted voluntarily in respect of a project which the developer believes is neither a Schedule 1 project, nor a Schedule 2 project that is likely to have significant environmental effects, it would be advisable for the avoidance of doubt to state that the ES is not being submitted for the purposes of the Town and Country Planning (Environmental Effects) Regulations 1988. If a developer voluntarily submits an ES expressly for the purposes of those Regulations, that is sufficient to determine that it is an application to which the Regulations shall apply.

Box 3.3 Factors determining desirability of a voluntary EA

Size and scale	-	Area
		Height
		Number of units
Sensitivity of site	-	SSSI
	-	Local/regional importance
	-	Archaeological/architectural interest
	-	Rare habitats
Adequacy of infrastructure	-	Draw on transport system
	-	Draw on drainage system
	-	Draw on public utilities
Nature of development	-	Experimental architecture
	-	Energy load/energy efficiency
	-	Impact on environmental media (i.e. discharges to atmosphere, water and land)

3.4 DETERMINING THE SCOPE OF THE ASSESSMENT

The first stage in the process of preparing an EA is to scan the range of parameters that might be affected by the proposal and scope around those aspects that will require more detailed consideration. Each proposal will generate a range of impacts. Initially, all potential impacts should be considered and evaluated. Those issues that are likely to be of significance should be identified in consultation with local bodies and interest groups. The identification of an issue at an early stage can save a great deal of subsequent expense, and will also assist in building a consensus relationship with the local population. After an initial overview, it will be feasible to omit from further evaluation those impacts that will clearly be so minor as to be of no consequence. However, before the decision is made to omit such potential impacts from further study, a statement should be produced setting out the justification for the decision, and it would be prudent to acknowledge in the EA itself that all issues have been considered notwithstanding that no impact has been identified.

Those impacts that are identified as having potential significance will be the subject of further research to establish the extent of that significance, and to identify, if necessary, any mitigating measures that can be incorporated into the development.

The establishment of local liaisons and consultations at this early stage can prove invaluable, not only in building up confidence in the developer's environmental assessment, but also as an important but often undocumented source of local knowledge.

Environmental impacts can occur at any stage in the built development proposal, whether it be site preparation, construction, operation or after use, and they may affect the world outside the development site or the users of the site itself. It is therefore necessary to assess the impacts that may occur at all the various stages in the life of any built development.

It may be necessary at this early stage to sound out the views of the local authority and other interested groups on what impacts any proposed project may have. This will in practice be an essential step where the proposed development is large scale. The end result of the scanning and scoping process will be a still fairly substantial list of items that require further investigation and consultation. There will generally also be more than one alternative scheme under consideration at this stage, each of which will require further testing and adjustment against the various environmental parameters.

The work done at this early stage will furthermore enable the developer or project planners to identify their next sources of data and possible consultees. It is important at this stage to consider possible timetabling of the EA process, particularly where monitoring may have to be carried out over several months (for example, certain habitats may have to be monitored throughout a 12-month cycle).

Once the issues have been identified, the next stage is to consider what data and what consultations may be required to analyse each aspect.

Table 3.2 sets out those consultees who may be useful sources of information and baseline data.

The consultation process will provide valuable local information, which may not be readily available from the publications and documentation to which the developer has access. Local interest groups and residents may thus identify particular environmental concerns at an early stage which will permit the development to be designed or located so as to minimise the environmental impact. The exercise of undertaking meaningful consultation may itself pre-empt or reduce the nature of the objections that may be raised by the interest groups at the planning stage. It is important that the consultation process should be at an early stage before the design and plans for the scheme become so advanced and fixed that modifications would be impracticable or extremely expensive.

As well as the importance of early consultation, the EA is a useful process in that it provides a systematic framework of assessment whereby the positive benefits of the scheme can also be seen, for example in the creation and protection of new habitats by the design, location and landscaping of the buildings.

Table 3.2 Potential consultees

Environmental characteristic	Example of appropriate data	Consultee
Traffic/transport	Road/rail capacity. Increased congestion	County council, Department of Transport, local planning authority, British Rail, other developers
Pollution	Land contamination, water pollution	Local planning authority, HMIP, NRA
Heritage	Architecture, archaeology	English Heritage, National Trust, Historic Buildings and Monuments Commission, Landmark Trust
Flora, fauna, birds	Habitat, species, rarity, location	English Nature, Countryside Council for Wales, RSPB, country wildlife trusts, CPRE
Existing centres	Effect of new development on vitality of existing centres. Regional growth projections	Local planning authority, development associations, Civic Trust, chamber of commerce, residents association
Flooding, drainage	Surface water and hydrology	National Rivers Authority
Infrastructure	Gas, water, electricity — overhead pylons	Gas/water/ electricity company, BT
Land use	Land use, population, soil/ rock type, geology	Local planning authority, Ordnance Survey, Ministry of Agriculture, British Geological Society
Amenity and leisure	Seasonal activity, local demand, catchment	Sporting associations, youth groups, other operators, tourist boards, local authority, Sports Council, CPRE, Countryside Commission
Planning issues	Green Belts/SSSIs, etc.	Planning authority, EN

3.5 POTENTIAL ENVIRONMENTAL IMPACT OF BUILT DEVELOPMENT

Where a formal EA is required, Schedule 3 to the 1988 Regulations specifies the information that must be included in the final Environmental Statement (see Chapter 2, Box 2.2).

It is clear from the requirements of Schedule 3 that the assessment of the impact of the built development upon the environment should consider the potential impact on each component part of the environment. Table 3.3 indicates those factors that may render an impact significant. Each of the major potential impacts will now be considered.

Table 3.3 Factors that may influence the significance of an impact

Impact	Significant factor
Location	Site in or close to sensitive location (eg SSSI, AONB, National park, Ramscar Site). Proposal would affect townscape/landscape. Proposed works in or close to listed building/conservation area. Development not sensitive to landscape context. Loss of playing fields/public open space.
Air pollution	Construction dust. Dust arising from demolition of former buildings (risk of asbestos contamination, etc.). Smoke from fires on clearance sites.
Water pollution	Urban site contains contaminated land. Surface/foul drainage problems. Disposal of waste materials from construction. Potential impact upon surface and underground water resources.
Ecology	Existing habitats of importance. Alternative schemes unable to preserve habitat. Interruption of wildlife corridors. De-watering of adjoining woodlands, etc. Tree Preservation Orders.
Noise and vibration	Site in close proximity to houses or offices. Particularly noise-sensitive neighbours including schools, hospitals, rest homes etc. Prolonged excessive noise impact during construction. Late-night noise during operation (e.g. bar/night club). Spectators at stadia. Traffic noise during construction and operation.
Transport	Increased road use, off-site parking. Proximity to existing transport infrastructure. Potential traffic hazard to nearby schools, bus routes etc. Transport of large construction plant and materials.
Visual aesthetics	Impact on the sky line; 'dominating' characteristic of development
Safety	Impact of development on main routes used by emergency services, either during or after construction
Infrastructure	Re-routeing of major public utilities, and secondary impact
Heritage/ geology	Ancient monuments in or near to site. Any listed buildings. Effect on landscape/townscape heritage. Impact on the skyline. Impact on geological sites.
Microclimate	Influence on wind direction; impact on temperature and convection etc.
Socio-economic	Impact on employment, recreation; secondary impact on other developments, etc.

3.5.1 Land contamination

The existing condition of the ground needs to be identified, i.e. what state the soil has been left in by past users such as factories and goods yards. This will almost certainly involve a general survey of the site. However, some useful information may be obtainable from the local authority concerning past planning permissions and other operating consents and an inspection of old plans and maps should be undertaken. The previous users themselves may also be a convenient guide as to the sort of contamination involved. Having identified which parts of the site require cleaning up, the question is whether to decontaminate the soil on site or to excavate and remove it. The former is generally the preferred method as it reduces lorry movements in the town and achieves more than simply shifting the problem to another location. However, removal of the contaminated soil is the more common method; in certain circumstances encapsulation on site may need to be considered.

The contamination of land can result in other environmental impacts off site, including:

- pollution of watercourses and groundwater
- air pollution (e.g. landfill gases, airborne asbestos fibres)
- increased traffic flows (if material required to be taken off site)
- impact on fauna and flora
- health and safety.

Care should be taken not only to deal with contamination from past uses of the site but also to prevent contamination arising during the construction of the development. The transport, delivery and storage of hazardous materials require to be assessed and procedures developed for use in the event of an emergency. Further guidance is currently being prepared (CIRIA, 1993a).

3.5.2 Air pollution: dust and smoke

The potential for the generation of dust is likely to be greatest during the pre-construction and construction phases of the development, although it may also be of importance during the operational phase, if the final land use is likely to lead to its creation (e.g. certain industrial manufacturing processes). In any event, the environmental impact assessment will almost certainly need to address the issue of dust, especially where the development is situated in an urban setting close to working or residential populations.

The assessment will need to consider potential dust sources, the likely impact on the site and its surroundings and possible mitigation measures (see Box 3.4).

Box 3.4 Potential dust sources

Pre-construction
Demolition of existing buildings
Removal of material from site
Excavation and division of services
Earthworks (where site reshaping necessary)

Construction
Transport to site of construction materials
Digging foundations
Erection of buildings
Bonfires

Operational
Process/activities
Transportation of final product

The assessment of each of the sources of dust will need to consider the potential environmental impact of that source (see Box 3.5). Sources in the middle of the site may have reduced potential impact as the majority of the dust will fall and settle within the construction site itself (although on large residential developments, phased over several years, new residents may be affected by ongoing works).

Box 3.5 Potential impacts: matters to consider

Location of site	How close is it to residential areas or working populations? Does it adjoin any sensitive land uses? If in a rural setting is it close to any sites of conservation importance, or sensitive flora, species etc?
Meteorological data	Wind speed and direction — where is the dust most likely to settle?
Constituents of dust	Could it be a health hazard to on-site personnel and/or local residents because of nature of materials used?

Having identified the sources of dust and assessed their potential impact, consideration needs to be given to measures which could be taken to mitigate that impact. These include:

- alternative construction methods
- suppression of dust by use of water bowsers on site
- wheelwashers to prevent dust being carried off site by vehicles
- keeping on-site and off-site roads clean
- sheeting lorries
- keeping windows closed.

Where the potential for dust generation is severe, the developer should consider whether monitoring is appropriate, both before development commences to assess ambient levels, and during construction and operational phases.

The developer must also have regard to relevant legislation and statutory controls relating to dust pollution, such as the Environmental Protection Act and the Clean Air Act.

Early consultations with the local environmental health officer are likely to be of assistance.

3.5.3 Water pollution

Many construction schemes, particularly those involving large-scale clearance of ground, extensive paving, the storage of materials and related activities pose a risk of water pollution.

During construction, the stripping of vegetation and topsoil can lead to large quantities of sediment-laden runoff from rainfall. The disturbance of contaminated ground will increase the pollutant load of this runoff. The storage on site of fuel and other materials also poses a potential pollution threat, particularly where work is undertaken in the vicinity of a watercourse, or in an area overlying an aquifer. Such risks are clearly dependent on the scale of construction and the sensitivity of the area.

After construction, the disposal of sewage will be subject to the control of the sewerage undertaker under the Water Industry Act. However, surface runoff can still create particular problems in terms of both quantity and quality. Where the runoff quantities are significant, the developer might find that balancing ponds are required. Such facilities often create an opportunity for enhancement through providing nature conservation or recreation activities, which can also enhance the value or appeal of the development. Where large vehicular parking or loading areas are provided, an additional pollution risk is introduced from spillages of oil and other materials. In this case, the developer will be required to provide suitable interceptors. Additional points for consideration include the following.

- Will demolition and site clearance involve potentially contaminative materials?
- If dust suppression by use of water is required, how will contaminated water be disposed of?
- How will surface water runoff be disposed of, both during and after construction?
- Is groundwater likely to percolate into excavations during construction and, if so, how will it be disposed of?
- Are any discharge consents required from the NRA?
- Are any local watercourses of particular ecological significance?
- Are there any potential drainage/flooding problems?
- Will the built development result in discharges of domestic or processing effluent?

3.5.4 Nature conservation: ecology and earth sciences

The more rural the setting, the greater the potential for an adverse ecological impact. In addressing this issue, the developer will first need to assess the ecological value of the site and its surroundings. Assistance in so doing is likely to be available from a variety of sources, and early consultation is likely to be cost-effective and time saving. The following organisations should be consulted (see also Table 2.2):

- English Nature
- Countryside Council for Wales
- Royal Society for Protection of Birds
- County Wildlife Trusts
- Council for the Protection of Rural England
- local authority
- local conservation organisations.

Consideration should first be given as to whether the site, or its surroundings, have been designated as having any ecological value (e.g. SSSI or feature or habitat of county significance) (see Chapter 2) and whether the site is the habitat of any protected species of flora, fauna or bird life (e.g. the EC Directive on the Conservation of Wild Birds and the Habitats and Species Directive). The site may not be of national importance but nevertheless may contain flora and fauna of regional or local importance.

In assessing the ecological impact of the development, consideration will need to be given to disruption during the construction phase of both on-site and off-site habitats and the potential for creating permanent, and possibly far-reaching ecological problems, such as by the severing of a wildlife corridor. In some cases an ecological survey, based on a full year's monitoring, will be appropriate. This will be particularly the case where a bird survey is required; the composition and total of the bird population will very enormously from season to season.

Where the potential for adverse ecological impact exists, the EA must look at mitigation measures such as:

- safeguarding habitats by sensitive design and siting of the development
- phasing construction so as to minimise disturbance e.g. by avoiding breeding or migration times
- re-siting or replacing habitats
- appropriate instruction of construction and operational employees.

For further guidance on vegetation surveys, the reader is referred to publications by English Nature (NCC, 1990; English Nature, 1994).

Consideration of potential impacts on the geological and geomorphological features of an area is also required. Almost a third of all SSSIs are designated for their 'earth science' interest. In addition to these sites of national importance, sites may also be important regionally or locally. In assessing the likely impact of development, a detailed survey of the area needs to be undertaken by an experienced geologist or geomorphologist. Direct physical damage may occur to exposures or land forms as well as indirect impacts caused by disruption of physical processes maintaining such features. Early consultation with the following organisations is likely to be cost-effective and time saving:

- English Nature
- Countryside Council for Wales
- Local authorities
- Local Regionally Important Geological Sites Groups (RIGs)
- Geological Association
- Geological Society.

Where impacts are identified, the EA must consider mitigation measures to:

- safeguard exposures/land forms/processes by sensitive design and siting of development
- enhance exposures currently found
- recreate exposures elsewhere.

Further details of the rationale behind earth science conservation and advice of practical design issues is covered in documents produced by English Nature (NCC, 1990; English Nature 1994).

3.5.5 Landscape

In cases where the development is situated in a rural location, the EA must have regard to the quality of the surrounding landscape and the potential impact of the new land use upon it. Initially, the developer will need to determine whether or not the site is in, or close to, a sensitive location, e.g. AONB, National Park. Sites not afforded statutory landscape significance may still be of local importance or be designated for protection within the local development plan. Where the development is likely to have an urbanising impact upon the landscape, care will need to be taken to minimise its effects as far as is possible by sensitive design and layout of the site and the use of bunding and screening (see also Section 3.5.9).

Further advice on this issue is provided by the Countryside Commission (1992).

3.5.6 Noise

The preparation for, and construction of, built development creates a large number of noise sources. The pre-construction phase has the potential to give rise to particularly high levels of noise, whether it be from the demolition of an existing building in an urban area or from the stripping of soil and preparation of the ground in a rural area. Where a building is to be demolished, each of the various options for effecting the demolition should be considered and the noise potential assessed. It is generally accepted that the public will tolerate a higher level of noise for the clearance of dereliction (in that they perceive a long-term advantage and benefit to the local environment) than is generally tolerated for new construction activity. The benefits of a higher noise level over a very short period (e.g. demolition by the controlled use of explosives) should be compared with a lower noise level over a longer period (e.g. where a building is dismantled so as to retain many of the features for incorporation in a subsequent development).

The choice of the appropriate means of demolition will not be determined by a simple reference to the noise level that will be generated. Nevertheless, the noise impact should be one of the considerations in choosing the method of demolition. The importance of dialogue in consultation prior to the choice of the demolition route cannot be overstated if objections and complaints are to be avoided at a later stage.

Since the Control of Pollution Act 1974 local authorities have been empowered to serve notices imposing requirements on the way in which construction, demolition and certain other specified works in relation to built development are to be carried out for the purpose of minimising noise. Such requirements can be specified before or after the work has commenced and this again is a reason for early consultation with the local authority. Noise generated by demolition and construction is heavily regulated and the main sources of control are listed in Box 3.6.

Box 3.6 Statutory controls over construction noise

Statutes
Control of Pollution Act 1974 – control by local authority
Health and Safety at Work Act 1974 – control by Health and Safety Executive
Land Compensation Act 1973 – where insulation works are required
Noise and Statutory Nuisances Act 1993 – noise from machinery in the streets

Statutory regulations
Noise Insulation Regulations 1975
Control of Noise (Measurement & Registers) Regulations 1976
Control of Noise (Codes of Practice for Construction & Open Sites) Order 1984
Construction Plant & Equipment (Harmonisation of Noise Emission Standards) Regulations 1985 & 1988

British Standards
BS 5228 – Noise Control on Construction and Open Sites
BS 4142: 1990 – Methods for Rating Industrial Noise Affecting Residential and Industrial Areas

Noise during the construction stage has the potential to be high and will often be regarded as intrusive as, by definition, the construction activity is a new noise source within the locality. It will be necessary to monitor and record the ambient noise levels in the local neighbourhood and then to agree with the planning authority the level by, and hours during which, the ambient can be exceeded during the construction phase. Some of the sources of noise during the construction phase are listed in Box 3.7.

Box 3.7 Sources of noise

Pre-construction
Demolition works
Ground preparation (e.g. soil stripping; clearance of rubble)
Excavation and diversion of services
Loading and unloading of lorries
Audible reversing warnings on vehicles
Mobile generators
Piling and excavation

Construction
Piling and foundation work
Various plant/compressors, cement mixers etc.
Operation of drills, hammers etc.
Audible reversing warnings on mobile plant.

3.5.7 Transport: traffic and congestion

Any development is likely to have some permanent effect on the use of local transportation facilities and it is usually the potential impact on the local highway network that will cause most concern and require detailed consideration in the EA, during both the construction and operational phases. The question is how to assess the size of this impact. Data may exist already within the local authority or Department of Transport for existing baseline figures in the area of the site, in other parts of the locality and generally for projections in the locality. In addition, extensive studies on similar projects in other regions will have been conducted and a good deal of time and expense could be saved by utilising these existing figures. These sources are well documented in the Department of the Environment publication *The Effects of Major Out of Town Retail Development* (HMSO, 1992). The developer will inevitably need to obtain more specific baseline data for the particular project. This will involve monitoring at a number of critical locations, such as junctions and proposed accesses. Ideally, the monitoring should cover a full range of seasonal fluctuations, particularly where the area concerned experiences marked seasonal variations in population.

Close liaison with the planning and highway authorities will generally be necessary in order to assimilate the project with the existing and future road infrastructure for the area. Where the potential impact of the development on the local highway network is likely to be great, this may involve linking the distributor roads with an existing or proposed ring road system and possibly making a contribution towards a new roadway. For a proposed retail scheme, for example, the local authority may wish to see the development linked to the provision of park and ride schemes.

The local authority will also require to be satisfied on the road safety aspects of the proposal, particularly with regard to neighbouring uses such as schools or housing.

Where a new highway is to be provided, or existing roads improved, there will be considerable advantages for the local community in terms of easing congestion, and the Environmental Statement can be used to emphasise this aspect to the public, who may feel that they have lost in other ways.

As well as traffic congestion, increased road use may create a noise impact, possibly many miles from the site. This may be compounded by nuisance from dust or mud during the construction phase. Impact can be anticipated by regular monitoring of public response at a range of locations throughout the planning phase.

Another problem associated with traffic is parking. This can be a major source of visual impact for some developments such as large retail schemes or leisure complexes and the car parks themselves will present considerable drainage and hydrology considerations.

The pre-operational stage of the development (i.e. site clearance and construction) is likely to generate heavy goods traffic, placing an additional burden upon the local network, although such increases in traffic movements will be temporary. However, the EA must not ignore construction traffic especially in the case of larger developments.

In assessing the long-term effect of the development on the highway the matters listed in Box 3.8 will require consideration.

Box 3.8 Impact on highway network material considerations

Type of development	What standard of highway network is required?
Site location	Urban or rural. Is it served by a good highway infrastructure?
Assessment of existing highway network	Current traffic flows and growth predictions Highway capacity and condition Speed restrictions
Traffic to be generated by development	Vehicle type Vehicle numbers Peak hour vehicle movement
Highway improvements	Are they necessary? Can they be carried out within the development site or will they require further land acquisition?
Safety aspects	Potential impact of increased traffic generation on sensitive locations e.g. schools, hospitals, residential areas Accident blackspots Speed controls
Local highway authority	Early consultation to canvass their views.

3.5.8 Public utilities

The impact of the development upon public utilities such as gas, water and electricity mains and water and sewerage services will need to be assessed from a variety of aspects, the relative importance of which will depend upon whether the development is sited in a rural or urban location. For example, in an urban setting the effect of construction upon existing services will be important whereas in a rural setting the more crucial question may be whether the utilities required to service the built development are already in place. Questions which need to be considered include:

- What public utilities currently exist on/close to the site?
- What is their precise route/location?
- Will the construction of the development require the diversion or relocation of any of the services?
- Are the utilities required by the development already in place and do they have the capacity to serve it?
- Will any improvement works be required and if so what is the likely impact of such works?

In order to deal with the above uses, extensive consultations are necessary with a variety of bodies. The principal service utilities are listed below:

- electricity generating company
- local electricity company
- local gas company
- telecommunications (BT, Mercury)
- local water companies and sewerage undertakers
- local authority.

3.5.9 Visual impact

During the construction phase, visual impact will result not only from the construction activities themselves but also from the removal of either the existing vegetation or previous buildings and other landscape features.

During the operational phase, visual impact will result both from the permanent features (the building, landscaping etc.) and from the activities associated with its use (parking of vehicles, etc.).

The final appearance of the development may well be an important factor. This will invariably be so in a rural setting and potentially equally so in an urban environment, particularly in a historic town centre setting. The subjective nature of visual intrusion means that the impact of a development is extremely difficult to assess. What may be attractive to one person will be 'a carbuncle' to another.

The methods used in the assessment will need careful consideration. They include:

- use of models and photomontages
- use of zones of visual intrusion
- selection of potential viewpoints into site.

Where the potential for adverse visual impact is great, because of the nature of the development, its prominence or the numbers of people likely to be affected, mitigation will be of particular importance. An attempt to define the geographical scope of the potential impact of a development is common to all methods of assessment, with the Zone of Visual Influence (ZVI) delineating the extent over which the development is likely to be visible. The constraints of cost, time and accessibility rarely enable all potentially affected locations to be visited to assess the visual impact. However, the use of computer programs to assist with the assessment is becoming increasingly common. These indicate areas of visibility as radii or a matrix of grid squares. Mitigation of visual impact can be achieved in many ways. These include:

- size and configuration of buildings, new roads, parking areas and proposed landscaping
- structure heights
- colour schemes
- landscaping — e.g. tree planting; construction of bunds
- site topography.

3.5.10 Socio-economic

The nature and extent of the socio-economic factors that an EA will have to consider will depend upon the type and location of the development. Examples of such factors are given in Box 3.9.

Box 3.9 Socio-economic effects

Effect upon local employment	Resulting directly from the development (construction and operational phases).
	Arising as an indirect result of the development (e.g. impact of out of town shopping malls on traditional retail centres)
Economic benefits	Will development increase local wealth and/or result in regional growth?
Effect upon local town centres	Retail developments may cause local high street dereliction; or relieve local traffic congestion
Amenity and leisure	Does the development create or reduce leisure facilities, public open space etc?
Effect upon surroundings	For example, is development removing an area of dereliction; bringing improvements to residential areas?

3.5.11 Heritage

Local heritage is highly valued. As part of the EA, the developer should therefore investigate whether or not the development will affect any ancient monuments, listed buildings, conservation areas, sites of archaeological or geological interest. Consultations with the following bodies would reveal whether and to what extent any such sites would be affected:

- English Heritage
- National Trust
- Historic Buildings and Monuments Commission
- Landmark Trust
- local authority – county archaeologist
- English Nature/CCW.

For listed buildings and ancient monuments, the developer would also need to have regard to the necessity of obtaining consents under the appropriate legislation.

Where the development will encroach upon a site of archaeological value, the developer should consider offering access prior to commencing the development to enable investigations to take place. The impact of building development on heritage and archaeology can often be significant.

3.5.12 Other impacts

The potential impacts referred to above are not intended to be an exhaustive list, but examples of those issues likely to be common to many developments.

Other issues that might also need to be addressed may include the effect upon the microclimate, where the location and size of the built development is such that it would have a significant impact upon the direction and strength of local winds.

The materials with which the building is to be constructed may also be subject to environmental assessment. The Building Research Establishment have published a report *BREEAM 1/90* to assist in the environmental assessment of new office designs and a further report *BREEAM 3/91* for the assessment of new dwelling houses (BRE, 1990; 1991). The stated purpose of the reports is to minimise the adverse effect of new buildings on the global and local environment while promoting a healthy indoor environment. Such global issues as greenhouse gases and ozone depletion are considered in this context.

The assessment may also extend to the environmental impact of the building upon the future users and occupiers of that building. The design and construction would thus be assessed in terms of risk and sick building syndrome, legionnaires' disease, etc. The Rosehaugh Guide entitled *Buildings and Health* (RIBA, 1992) is a useful publication in this area and contains a checklist for 'benign building design'.

CIRIA has produced a useful review of environmental issues and initiatives relevant to the construction industry, through their Construction Industry Environmental Forum (CIRIA 1993b).

The Department of Energy has also produced a wide variety of publications on the conservation and efficient use of energy in the design construction and management of buildings (DoE, 1992).

3.6 ENVIRONMENTAL ASSESSMENT IN PRACTICE: CASE STUDIES

The purpose of this section is to give a flavour of recent examples of the assessment of the environmental impact of built developments in both urban and rural settings. The developments for which full EAs have been carried out are by their nature large but demonstrate the interplay between the various categories of environmental impact. The mitigation of one impact may well lead to an increase in the impact of another environmental factor. The final design of a development should thus be assessed as a whole as well as in terms of its constituent environmental impacts. See also Watson (1992).

3.6.1 Case study 1

The first case study is a residential development of over 5,000 dwellings on 600 acres of formally derelict chalk quarries. Residential schemes are, by their very nature, people-orientated and have predetermined requirements. The objectives for the development of the new settlement were:

- to repair the environmental damage of the past
- to provide land for development
- to retain and enhance existing features for public enjoyment
- to create a community with a sense of belonging
- to provide recreation and leisure facilities for the residents of the area
- to facilitate improved communications.

Part of the site, which contained two flooded quarries with dominating vertical chalk cliffs, posed conflicting requirements between safety and existing landscape. This was resolved by the preservation of the cliffs and the provision of a safety zone along the bottom of the cliffs to which public access was to be prohibited.

A new community of this size produces a substantial increase in traffic, particularly in the morning peak hour. Three new main access points providing connections for the local trunk and motorway network were constructed and, where appropriate, houses constructed in the vicinity of the new highways were designed so as to incorporate the necessary noise insulation. The internal road pattern was dictated by the identified ecological constraints including Sites of Special Scientific Interest, woodlands and nesting areas. Further, a network of 'greenways' was provided for linking the whole development for pedestrians and cyclists on landscaped routes away from the roads.

One of the main environmental impacts of built development in a rural setting is the potential impact upon the ecology. The wealth of flora and fauna on the site was enormous in terms both of scale and variety. Early consultations with the (then) Nature Conservancy Council identified two ecological sites of special scientific interest, one of which related to flora and the other to a rare spider. Concerns were also expressed as to the potential impact upon the established bird habitats. Early discussions with the local authority planning officers drew attention to the value of the woodland hedgerows and the potential existence of badgers within the site. Early access to such information enabled clear and strong ecological constraints to be imposed during the formulation of the layout of the proposed development. It also enabled the subsequent ecological surveys to be focused and this in turn revealed more species of fauna and flora that would require to be protected.

A detailed ecological study of the site recommended a retention of identified high-value areas, combined with the deliberate creation of ecologically appropriate habitats as part of the new landscaping for the project. The pressures on wildlife during the construction period would be particularly great, and to mitigate these pressures a safe approach was used which also had regard to the breeding seasons of the identified species. Concern was also voiced by the ecologists of the potential effects of contaminants, particularly silt from highway runoff, on the aquatic life in the retained lakes. To mitigate this, settlement ponds were designed as an integral part of the drainage scheme.

The overall design of the development, particularly in terms of layout, required a careful balance to be struck between the colonisation of the site by man and the colonisation by nature that had already occurred. This balance was necessary not only in the design and construction but also in the future operation of the development by preserving certain habitats as 'no go' areas for local residents. This was achieved both by the routeing of footpaths and highways and also by the retention of physical barriers (e.g. water areas) to avoid disturbance of important habitats by local residents, and by education, e.g. through local schools and the provision of information boards on the site.

3.6.2 Case study 2

The second case study involves a mixed development in an urban setting. The substantial site contained a mixture of pre-existing uses including industrial retail and residential. While the impact of the development on ecology had to be considered, the impact was slight, particularly compared with the previous case study. A large-scale development in an urban area rather raised concern of the potential and environmental impacts upon man with particular reference to noise, dust and traffic congestion. The identification of areas of contamination within the site at an early stage enabled the development to be designed and phased so that the most contaminated material could be taken off site to a licensed waste disposal facility, while the less contaminated material could be encapsulated on those areas of the site that would be used in the future for car parking etc. The identification of the contaminated materials at an early stage followed by early discussions with the National Rivers Authority, waste regulatory authority and local planning authority all ensured that the past effects of the previous use of the site could be dealt with in a considered and programmed way.

Monitoring of ambient levels in terms both of air pollution and noise was undertaken around the site, in cooperation with the local environmental health department, prior to any form of development on the site. Appropriate limits and hours of working were then agreed with the local planning officers and this enabled a relatively trouble-free construction period whereby the developer, planning authority and local residents were each aware of the noise conditions that would apply during the construction phase.

The early involvement of local conservation groups, civic groups etc. again proved to be a great value to both the groups and the developer alike as their concerns could be addressed at an early stage in the formulation of the design and landscaping of the site.

Again, early consultation with the public utilities proved to be extremely important, not only to ensure that the new development would be adequately serviced, but also to identify and preserve the need for the existing services underlying the site. The early consultation enabled the water company's own programme for the replacement of infrastructure to be brought forward, with each party bearing a contribution of the costs. The advantages of early consultation cannot therefore be overemphasised.

3.7 REFERENCES

Directives

Department of the Environment Circular 1/92 — Planning Controls over Sites of Special Scientific Interest

European Community Directive — Council Directive on the assessment of the effects of certain public and private projects on the environment (85/337/EEC)

European Community Directive — Council Directive on the conservation of wild birds (79/409/EEC)

European Community Directive — Council Directive on the conservation of natural habitats and of wild fauna and flora (92/43/EEC)

Legislation

National Parks and Access to the Countryside Act 1949

Land Compensation Act 1973

Control of Pollution Act 1974

Health and Safety at Work etc. Act 1974

Ancient Monuments and Archaeological Areas Act 1979

Highways Act 1980

Wildlife and Countryside Act 1981

Derelict Land Act 1982

Wildlife and Countryside (Amendment) Act 1985

Environmental Protection Act 1990

Planning (Hazardous Substances) Act 1990

Town and Country Planning Act 1990

Land Drainage Act 1991

Planning and Compensation Act 1991

Water Industry Act 1991

Water Resources Act 1991

Protection of Badgers Act 1992

Clean Air Act 1993

Noise and Statutory Nuisances Act 1993

Regulation

Noise Insulation Regulations 1975

Control of Noise (Measurement and Registers) Regulations 1976 (SI 1976/37)

Control of Noise (Codes of Practice for Construction and Open Sites) Order 1984 (SI 1984/1992)

Construction Plant and Equipment (Harmonisation of Noise Emission Standards) Regulations 1985 (SI 1985/1968)

Construction Plant and Equipment (Harmonisation of Noise Emission Standards) Regulations 1988 (SI 1985/361)

Town and Country Planning (Assessment of Environmental Effects) Regulations 1988 (SI 1988/1199)

Town and Country Planning (Assessment of Environmental Effects) (Amendment) Regulations 1990 (S.I. 1990/367)

Town and Country Planning (Assessment of Environmental Effects) (Amendment) Regulations 1992 (SI 1992/1494)

Department of the Environment Circular No 15/88 – Town and Country Planning (Assessment of Environmental Effects) Regulations 1988

Welsh Office Circular No 23/88 – Town and Country Planning (Assessment of Environmental Effects) Regulations 1988

Technical

BSI (1984–86)
BS 5228 Noise control on construction and open sites
British Standards Institution (Milton Keynes)

BSI 1988
BS 4142 1990 – Methods for rating industrial noise affecting residential and industrial areas
British Standards Institution (Milton Keynes)

BRE (1990)
BREEAM 1/90 – an environmental assessment for new office designs
Building Research Establishment, Watford

BRE (1991)
BREEAM 3/91 – an environmental assessment for new dwellinghouses
Building Research Establishment, Watford

CIRIA (1993a)
Contaminated land
CIRIA (London)

(CIRIA 1993b)
Environmental Issues in Construction
CIRIA (London)

Countryside Commission (1992)
Environmental Assessment — the treatment of landscape and countryside recreation issues CCP 326
Countryside Commission, Cheltenham, Glos.

DoE (1992)
Building Management: Environmental Action Guide
Department of the Environment

English Nature (1994 in press)
Nature Conservation in Environmental Assessment
English Nature, Peterborough

HMSO (1989)
Environmental Assessment: A guide to the procedures
Department of the Environment and Welsh Office

HMSO (1992)
The Effects of Major Out of Town Retail Development
Department of the Environment

NCC (1990)
Handbook for Phase 1 Habitat Survey
Nature Conservancy Council

RIBA (1992)
Buildings and Health The Rosehaugh Guide to the Design, Construction, Use and Management of Buildings
RIBA Publications

Watson, S. (1992)
Environmental aspects of residential development
IWEM, South-Eastern Branch Paper, November 1990

CIRIA Research Project 424

Environmental Assessment

A guide to the identification and mitigation of environmental issues in construction schemes

4. River and coastal engineering

Jan Brooke, Manager, Posford Duvivier Environment

4 River and coastal engineering

4.1 TYPE OF SCHEME CONSIDERED

This chapter deals primarily with the environmental impacts of engineering projects undertaken at the water's edge. It focuses on two types of river and coastal engineering schemes:

- flood defence and coast protection works
- port, harbour and marina development and redevelopment projects.

The various types of structure or non-structural solution that are considered are listed in Table 4.1. The approach adopted is, however, common to most schemes undertaken at the land−water interface and, as such, the chapter should also be generally appropriate to other projects. River and coastal control structures, barriers and flood defence barrages are therefore briefly considered. Projects excluded from consideration in this chapter include reservoirs and dams, agricultural and urban drainage schemes, irrigation projects and pollution control measures. Many of these are dealt with in other chapters.

4.2 KEY STANDARDS AND LEGISLATION (1992)

4.2.1 The need for Environmental Assessment

Many activities, often potentially conflicting, compete for space and facilities at the water's edge. The natural environment, in particular, has suffered as a result of development pressure. Today, the sensitivity of the water's edge environment is widely acknowledged and many agencies do now carry out environmental investigations prior to undertaking river or coastal engineering works.

The process of EA (i.e. the identification, evaluation and, if appropriate, mitigation of environmental impacts) should be applied to all water's edge developments. The first steps in this process (see Section 4.3) enable potentially significant impacts to be identified. In some cases a full EA will then need to be undertaken. This is likely to lead to the production of an Environmental Statement (ES) under the terms of one or more of a suite of Statutory Instruments (SI) which implemented EC Directive EEC/85/337 (on the assessment of the effects of certain projects on the environment) from July 1988 onwards. In other cases the preliminary evaluation will demonstrate that there are unlikely to be any significant environmental impacts associated with the proposed project. Given such an outcome it may be that no further EA work is necessary, but the investigation must nevertheless be documented (e.g. the 'written justification' produced by National Rivers Authority Thames Region for projects not requiring an ES). This document will provide the evidence, should it be required, that potential environmental impacts have been considered.

Many projects, for example marinas and some harbour developments, are subject to planning controls and need to be granted planning permission before proceeding. These projects might require a full EA under the terms of SI 1988/1199, The Town and Country Planning (Assessment of Environmental Effects) Regulations 1988. Where planning permission is not a requirement of the development (e.g. works carried out under the terms of a General Development Order), an EA may nevertheless be required under one or more of the following:

SI 1988/1217   The Land Drainage Improvement Works (Assessment of Environmental Effects) Regulations 1988
SI 1988/1336   The Harbour Works (Assessment of Environmental Effects) Regulations 1988
SI 1989/424   The Harbour Works (Assessment of Environmental Effects) (No.2) Regulations 1989.

Table 4.1 Examples of river and coastal engineering projects

Scheme type	Typical promoter	Type of works	Type of structure/solution
Flood defences[1] (SI 1988/1217; SI 1988/1199)[2]	National Rivers Authority; local authority; internal drainage board; riparian owner	Sea defences	Sea walls Beach nourishment Groynes Offshore breakwaters Embankments Revetments Vertical walls (e.g. piling)
		River works	Channel improvements Embankments Weirs, sluices and flow Control structures Flood storage reservoirs Flood diversion channels Culverts
Coast protection[1,3]	Local authority; private owner		Sea walls Beach nourishment Groynes Offshore breakwaters Revetments Embankments Vertical walls Drainage systems
Ports and harbours (SI 1988/1336; SI 1989/424[2])	Port authority; harbour commissioners; harbour conservancy; local authority; private developer	Expansion	Reclamation (land claim) Dredging Quays/wharfs/slipways Piers/jetties/dolphins Roll-on roll-off ramps Infrastructure Buildings
		Redevelopment	Buildings Infrastructure Marinas (see below)
Marinas (SI 1988/1199)[2,4]	Private developer; local authority		Pontoons Moorings Infrastructure Buildings Dredging

Notes:
[1] In England and Wales, flood defences are promoted under the Land Drainage and Water Resources Acts 1991 to mitigate flooding of low-lying land. Such flood defences are frequently maintained by the NRA although many are either in private ownership or are the responsibility of the local authority. Internal drainage boards also have land drainage responsibilities. Coast protection refers to schemes promoted under the terms of the Coast Protection Act 1949 to mitigate coastal erosion. Coast protection is primarily the responsibility of the local authority, but may be undertaken by private landowners. In both cases (i.e. flood defence and coast protection) grant aid may be payable on capital projects from the Ministry of Agriculture, Fisheries and Food (Welsh Office in Wales).
[2] Statutory Instrument under which EA might be required.
[3] Coast protection works are currently being considered as an addition to Annex II of EC Directive EEC/85/337, making them subject to SI1199.
[4] SI1199 applies only to developments above mean low water.

In common with EAs for all projects, the circumstances under which an EA might be required vary significantly from case to case. On some projects an EA is mandatory; in most cases for river and coastal engineering, however, it is discretionary. The only river and coastal engineering project that falls in Annex I of EC Directive EEC/85/337 (i.e. projects on which EA is mandatory) is as follows:

- trading ports, and also inland waterways and ports for inland waterway traffic that permit the passage of vessels over 1,350 tonnes.

A large number of projects fall under Annex II (i.e. projects on which EA is discretionary). These include:

- the construction of .... harbours, including fishing harbours .... (projects not listed in Annex I)
- canalisation or flood relief works
- yacht marinas.

At the time of writing (August 1992) coast protection works are not included in either Annex. Consideration is, however, being given to including such works as an Annex II project. This could be achieved using Section 71A of the Town and Country Planning Act (inserted by Section 15 of the 1991 Planning and Compensation Act).

The production of an EA under Annex II may be volunteered by a developer or it could be required/requested by any one of a number of authorities. Such authorities include:

- the local planning (or highways) authority
- the Ministry of Agriculture, Fisheries and Food
- the Secretary of State for:
    - Transport (harbours, docks, piers and ferries)
    - Agriculture (land drainage)
    - Environment (town and country planning)
    - Wales (developments in Wales).

For projects on which it is not intended that an EA be prepared, various agencies such as English Nature, the Countryside Commission, the Countryside Council for Wales and the National Rivers Authority, among others, can make representations to the appropriate authority (or, if necessary, the Secretary of State) and request that an EA be prepared. For example, the Land Drainage SI (1988/1217) dictates that a drainage body proposing to carry out any improvement works shall announce their intentions in at least two local newspapers, thus giving agencies such as those mentioned above the chance to make representations.

Although the Department of the Environment, National Rivers Authority, English Nature and others are currently (August 1992) reviewing the guidance available to developers, formal guidance on whether or not an EA is required (e.g. for projects such as harbour developments and coast protection works) is still limited. Developers are therefore advised to check with the appropriate authority at an early stage in the feasibility studies, to establish whether or not an EA will be required for a particular project. Appropriate authorities include those issuing planning permission, licences and consents.

The existing guidance for EA on projects requiring planning permission is given in DoE circular 15/88 (Welsh Office 23/88) as follows:

> In general terms the Secretary of State's view is that environmental assessment will be needed for Annex 2 projects in three main types of case:
>
> i.   for major projects which are of more than local importance
>
> ii. occasionally for projects on a smaller scale which are proposed for particularly sensitive or vulnerable locations
>
> iii. in a small number of cases, for projects with unusually complex and potentially adverse environmental effects, where expert and detailed analysis of those effects would be desirable and would be relevant to the issue of principle as to whether or not the development should be permitted.

This guidance note also emphasises that the basic test of the need for environmental assessment in a particular case is the likelihood of significant environmental effects, and not the amount of opposition or controversy to which a project gives rise, except to the extent that the substance of opponents' arguments indicates that there may be significant environmental issues.

Finally, the Department of Transport (personal communication, 1990) comment that, subject to the evolution of common practice for port and harbour developments, the EC Directive is considered to target:

- major developments (completely new port facilities)
- developments likely to have significant environmental effects on people, wildlife, landscape, the heritage, etc, by virtue of their nature, size, location, etc.

4.2.2 Other relevant legislation

In addition to the Statutory Instruments introducing EA on certain projects, river and coastal engineering projects are covered by a wide range of other national, European and international legal requirements in the form of Acts, Conventions and Directives. Some of these are of particular importance to the EA process because they place constraints upon development or set standards that must be met by the development. Others, for example Planning Policy Guidance Note 12 (Development Plans and Regional Planning Guidance) and Planning Policy Guidance Note 20 on Coastal Planning offer overall policy guidance to planning authorities and other agencies. The main UK Acts of relevance to the EA process are shown in Table 4.2. Table 4.3 outlines the various international controls pertinent to river and coastal engineering developments, while Table 4.4 lists the European Directives relevant to the EA process.

Table 4.2 Summary of UK legislation relevant to the EA process for river and coastal engineering

Date	Act	Purpose/Relevance
1949	National Parks and Access to the Countryside Act	National parks; nature conservation: public rights of way; access to open country
1949	Coast Protection Act	General powers of coast protection authorities; control structures below low water mark
1975	Salmon and Freshwater Fisheries Act	Regulation of inland fisheries; salmon and sea trout up to 6 miles
1981	Wildlife and Countryside Act	Nature conservation; SSSI designation; protection of rare species
1985	Food and Environment Protection Act	Licences for construction works and dumping at sea; control over pollution of coastal waters
1985	Wildlife and Countryside (Amendment) Act	SSSI notification
1990	Environmental Protection Act	Creating English Nature and Countryside Council for Wales; limiting Countryside Commission activities to England. Introduction of Integrated Pollution Control. Certain amendments to the Food and Environment Act 1985 on licensing pollution which is extended to 'UK controlled waters'
1990	Town and Country Planning Act	Requirement for planning permission
1991	Water Resources Act	General duties of the National Rivers Authority; primary source of pollution control in relation to water; NRA's environmental and recreational duties; flood defence
1991	Land Drainage Act 1991	Consolidation of enactments dealing with land drainage; Powers of Drainage Boards and Local Authorities
1991	The Water Industry Act	General duties of water undertakers; sets standards in relation to the quality and sufficiency of water supply
1991	Water Consolidation (Consequential Provisions) Act 1991	Amendments and repeals in relation to the new Water Act 1991
1991	Planning and Compensation Act	Assessment of environmental effects

Table 4.3 International legislation relevant to the EA process for river and coastal engineering

Date	Convention/Directive	Purpose
1958 /1982	United Nations Convention on the Law of the Sea (UNCLOS)	Control, prevention and reduction of marine pollution
1971	Ramsar Convention on the Protection of Wetlands of International Importance	Designation of internationally recognised sites
1972	Oslo Convention for the Prevention of Marine Pollution by Dumping from Ships and Aircraft	Control of marine pollution by dumping and incineration
1972	London Convention for the Prevention of Marine Pollution by Dumping of Wastes and other Matter	Control of marine pollution by dumping and incineration
1973	International Convention for the Prevention of Pollution from Ships (MARPOL)	Annex I controls oil pollution; Annex II controls noxious liquids; Annex V controls the disposal of garbage
1979	Berne Convention on the Conservation of European Wildlife and Natural Habitats	Mandatory requirements to conserve wild flora and fauna and their natural habitats, giving particular emphasis to endangered and vulnerable species
1979	Bonn Convention on the Conservation of Migratory Species of Wild Animals	Strict protection for certain species in danger of extinction throughout all or a significant portion of their range; agreements for management of other species

Table 4.4 Summary of European Directives relevant to the EA process for river and coastal engineering

Date	Directive	Purpose
1976	Directive on pollution caused by certain dangerous substances discharged into the aquatic environment (76/464/EEC)	Seeks to protect all inland surface waters, internal coastal waters and territorial waters from pollution
1976	Quality of Bathing Waters (76/160/EEC)	Sets quality requirements for bathing waters
1978	Quality of Freshwater for Fish (78/659/EEC)	Deals with the quality of water required to support salmonoid and cyprinid species of fish
1979	Quality of Shellfish Waters (79/923/EEC)	Sets quality requirements for shellfish waters
1979	Wild Birds Directive (79/409/EEC)	Setting up of Special Protection Areas (SPA) for birds
1980	Directive on the protection of groundwater against pollution caused by certain dangerous substances (80/68/EEC)	Seeks to prevent pollution of groundwater and check or monitor compliance with any conditions imposed
1985	Environmental Assessment (85/337/EEC)	Assessment of the effects of certain projects on the environment
1993	Habitats and Species Directive: Modified Draft expected 1992	Defining specially protected habitats

4.3 PRINCIPAL ACTIVITIES AND MAIN ENVIRONMENTAL EFFECTS

4.3.1 The scoping exercise

Depending on the scale of the proposed development, some combination of site visits, literature and policy reviews, an initial data-collection exercise and a first round of consultation with interested parties should provide an indication of the type of issues which are likely to be important on a scheme-specific basis. This process will highlight not only genuine environmental concerns but also perceived problems (possibly due to a lack of understanding) or politically motivated issues, all of which will need to be recognised and dealt with during the course of the study.

A major project, such as a new flood diversion channel or a significant port expansion, is likely to require a thorough, detailed preliminary review. For a smaller project, such as the replacement of a culvert under a road, a discussion involving all interested parties may suffice. Maintenance works should be subject to the same sort of preliminary evaluation process as capital works because, in some cases, the former can have significant environmental impacts.

This issue identification process and preliminary evaluation is known as scoping and the purpose of the exercise is to identify the key issues to be addressed by the EA. This in turn guides the allocation of resources and ensures that they are correctly targeted for the remainder of the study. There is an urgent requirement with all EA work to direct resources towards the investigation of those environmental characteristics that may be significantly affected by the development, *not* simply the environmental characteristics in which those responsible for the EA happen to have expertise. For example, there is no need to spend money investigating operational traffic issues if it can be demonstrated that the proposed port development will not lead to any immediate or longer-term change in the status quo.

The scoping exercise should also serve to identify possible environmental impacts, the investigation of which may require an input from specialist advisors. Reclamation works might, for example, cause a change in the hydrological or coastal regime and physical modelling might be required to identify and mitigate against possible adverse downstream impacts.

Mathematical modelling might be required to ascertain the likely dispersal characteristics of pollutants entering a watercourse. Geotechnical advice might be sought on slope stability following capital dredging works; or a marine biological survey might be required in the absence of existing information on benthic flora and fauna. In all cases, however, the scale of the investigation should be appropriate to the importance of the issue.

Finally, the scoping exercise, by determining the nature, extent and significance of environmental interest and hence the sensitivity of the location, should provide a good, if not definitive, indication of whether or not it will be necessary to prepare a formal Environmental Statement.

4.3.2 Baseline data requirements: project activities

The first stage in the EA process, which should ideally be undertaken at the pre-feasibility stage of a project while all options (including doing nothing) are still under consideration, is the definition of the problem and the establishment of an appropriate database (see Figure 4.1). In order to comply with Annex 3 of EC Directive 85/337/EEC, this should comprise all the data necessary to identify and assess the main effects that the proposal is likely to have on the environment. Major projects, or those in particularly sensitive locations, are likely to have the more extensive data requirements both in terms of scope and diversity. It is important, however, that the data for both small and large projects should, as far as practicable, be scientifically and statistically valid (e.g. an adequate number of samples).

All projects will require a (site-specific) description of both the construction and the operational phases of the proposed development, together with a description of the aspects of the existing environment likely to be significantly affected by that project. The former may include activities such as dredging; extraction or importing of materials (terrestrial or marine); site works such as piling, bund construction, etc; placing of fill (e.g. for land claim) and generation of land- and water-based traffic. In common with most development projects, the activities associated with the construction phase of river and coastal engineering works can be very different from those associated with the operational phase of the development. Sheet piling activities, for example, can cause disturbance during the construction of sea defence works, but noise is not usually a significant impact associated with the operational phase. Capital dredging activities can potentially have more serious implications for inter-tidal nature conservation interests than subsequent maintenance dredging. Water pollution or land contamination, however, are more likely to be associated with the operational phase of (port and harbour) developments than with their construction phase. Excavation works on the beach might damage geological or archaeological exposures, but coastal defence solutions involving beach nourishment can, conversely, provide such features with a better degree of protection. Further, the long-term operational effects of a flood defence or coast protection project may appear somewhat innocuous in comparison with those associated with port, harbour and marina developments, with the latter supporting potentially noisy, polluting or otherwise disturbing activities.

Table 4.5 describes some of the major activities associated with each of these categories of works, and can be used to identify cases where a particular activity might regularly or occasionally be expected to result in a potentially significant impact. Table 4.5 also briefly reviews the possible activities associated with the decommissioning or abandonment of a particular structure. Overall, when identifying activities pertinent to a particular project, it should be anticipated that most types of development will lead to environmental impacts associated with activities during the construction phase. During the operational phase, however, there are significantly more activities associated with port and harbour developments than with flood defence or coast protection works.

Table 4.5 Activities that may result in potentially significant impacts

Activity	Examples
Construction phase	
Site works	Land clearance; earthworks; placing of materials
Capital dredging	Beach nourishment materials; new berth
Ancillary works	Temporary access; power supplies; water etc.
Raw materials extraction	Rock armour; aggregates; gravel (impact of materials extraction on source)
Transport of raw materials	Rock, concrete, steel structures etc.
Transport of employees	Construction personnel
Immigration	Contractors; temporary workforce
Employment	Employment of local people
Local expenditure	By developer on services and materials/employees on goods and services
Lighting	Construction site
Vibration	Pile driving; traffic; demolition
Noise	Transport; pile driving; demolition
Dust and particulates	Earth moving; demolition; concrete batching
Gaseous emissions	Exhaust emissions
Aqueous discharges	Pollution incidents
Dredged material disposal	Capital dredging
Solid waste disposal	Materials on site; old structures; contaminated land
Accidents/hazards	Collision; fire; structural failure; landslip
Operational phase	
Structures	Seawalls; embankments; jetties; ramps; cranes etc.
Maintenance dredging	Approach channel; vegetation removal
Raw materials extraction	Beach recharge; rock armour
Transport of raw materials	Imported and exported goods; aggregates
Transport of products	Imported and exported goods
Transport of employees	Operational staff
Immigration	Permanent workforce
Employment	Long-term job creation/loss
Local expenditure	By operator on materials/employees on goods and services
Lighting	High mast lighting of storage areas; nighttime operations
Vibration	Traffic
Noise	Port operations; traffic
Odours	Cargo handling; sewage
Dust and particulates	Bulk cargo handling
Gaseous emissions	Exhaust emission

continued/...

Table 4.5 (continued)

Activity	Examples
Operational phase	
Aqueous discharges	Cargo-handling incidents; bilge waters; oils; surface run off; collision
Dredged material disposal	Maintenance dredging
Solid waste disposal	By-products; litter; ship's waste
Accidents/hazards	Collision; fire; explosion; structural failure
Post-operation	
Cessation of activity	Employment; transport
Deterioration of structures	Degradation; collapse
Removal of structures	Demolition
Contamination	Long-term pollution of soil, water or sediment
Disposal of residues	Contaminated land

4.3.3 Common environmental impacts associated with river and coastal engineering works

Using a proposed oil berth development as a case study, Figure 4.2 demonstrates those areas of interaction that might be expected to give rise to environmental impacts when river and coastal engineering projects are being assessed. On the whole, as indicated in Section 4.3.2, port and harbour developments should be expected to have a higher frequency of interaction with the various environmental parameters than flood defence and coast protection projects, largely because of the passive nature of the latter during the operational phase. It is not possible within this document to explain every one of these potential interactions and Section 4.4 therefore concentrates on some of those impacts that previous environmental assessments have highlighted as occurring most frequently and/or being of greatest potential significance. It should be stressed, however, that each scheme will have different characteristics and different impacts. The contents of this section should therefore provide nothing more than an indication of what might be expected.

The impact identification and evaluation process depends, as discussed earlier, on high-quality baseline data and a well-founded understanding of both the existing environment and the proposed project. Potential direct and indirect (or knock-on) consequences of the project can then be identified. Some of the impacts of river and coastal engineering works can be predicted with reasonable accuracy. The number of jobs likely to be created by the construction work or the number of trips required to transport materials to site can be quantified relatively easily and the significance of such impacts can therefore be assessed in local, regional or national terms. In other cases, however, particularly at the coast, potential impacts are relatively more difficult to evaluate, as much depends on the vagaries of the weather and its impact on winds, waves and tides. Computer, mathematical, hydraulic and physical modelling are now commonplace in river and coastal engineering. Models can, however, be expensive, and models developed for engineering purposes may not be suitable for answering environmental questions. Similarly, care should be taken to ensure that the predictions of these models are verified throughout not only the construction phase of a project but also its operational phase.

Although physical and/or computer modelling can be used to predict, for example, the likely impacts of a new structure on patterns of downstream erosion and deposition, monitoring will nevertheless be required to check that no long-term damage is being caused, to fisheries or nature conservation interests for instance. Similarly, many adverse impacts are, in theory at least, preventable and prevention is invariably preferable to remedial action. Again, long-term monitoring should be undertaken, for example to ensure that apparently minor leakages or spills at ports and marinas are being correctly dealt with and prevented from entering the water body, and that the various safety systems (e.g. lighting, communication) that are critical in ensuring safe navigation when floating plant is in use are functioning effectively.

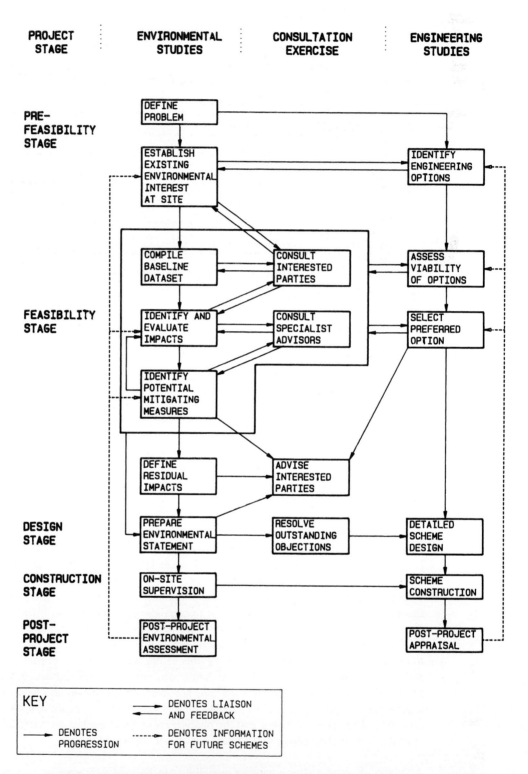

Figure 4.1 *The environmental assessment process*

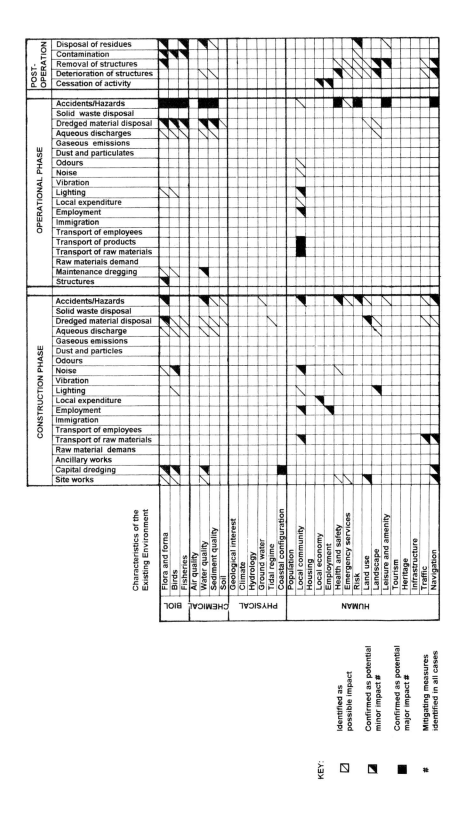

Figure 4.2 *Environmental impacts of proposed oil berth development*

4.3.4 Baseline data requirements: environmental parameters

The environmental characteristics that may be affected to a greater or lesser extent by river and coastal engineering schemes are generally common to all the types of development under consideration in this chapter. These characteristics are listed in Table 4.6, along with a checklist of the common information requirements for the preparation of an adequate baseline dataset. It should be noted, however, that this checklist is not definitive.

4.3.5 The role of consultation in the data-collection process

Much of the information required for any particular EA is likely to exist already: in published reports, in files in someone's office, or in somebody's head. Within reason, every effort should be made to track down these existing data because, even if the information does not exist or does not exist in the required form, the investigation will almost always reveal other interesting and relevant facts about, for example, the development site. Making maximum use of existing data also reduces the overall cost of the study to the developer.

This does not mean, however, that data of a substandard quality should be accepted. High-quality information is one of the most powerful tools in EA. A literature search will help to provide essential background data. A review of historic and current OS maps and aerial photographs will help to establish whether or not there have been any significant changes in the physical or topographical features of the site (e.g. progressive coastal erosion or its abrupt cessation; the development of river meanders; significant land claim, etc.). An appraisal of existing databases might also be appropriate, although this can be expensive and relevant databases should therefore be carefully identified and targeted.

Details about the development options under consideration should be available from the developer or the engineers on the study team. Close liaison between the engineering team and the environmental team is essential throughout the EA process in order to ensure that concerns are communicated and questions raised at the earliest possible stage.

In terms of collecting information about the existing environment, consultation plays a critical role in the EA process. A thorough programme of consultation not only enables those undertaking the EA to draw on a wide range of experience and local knowledge, but also helps in identifying potential impacts. The statutory need to consult varies according to the type of development being undertaken. Those agencies that are considered to be statutory consultees under Regulation 8 of SI 1988/1199 are shown in the DoE/Welsh Office publication *Environmental Assessment: A Guide to the Procedures* (HMSO 1989).

The Statutory Instruments implementing the Land Drainage and Harbour Works Regulations are, however, somewhat non-committal on the issue of statutory consultees. SI 1217/1988 deals with the requirement for any organisation to make available to the drainage body all relevant information in their possession, for a reasonable charge if appropriate. It also states that copies of the notice advertising that an Environmental Statement has been prepared shall be supplied to the Nature Conservancy Council (now English Nature and the Countryside Council for Wales), the Countryside Commission and any other public authority, statutory body or organisation that appears to have an interest in the matter. No mention is made, however, of the need to consult with such bodies during the preparation of the EA.

The Department of Transport (personal communication, 1990) report, in respect of the harbour works SIs, that the 'bodies with environmental responsibilities... to be supplied with copies of information on environmental effects and to be consulted are the local authorities, the Countryside Commission, the Nature Conservancy, and Her Majesty's Inspectorate of Pollution'.

Table 4.6 Baseline data requirements

Environmental characteristics	Examples of appropriate data
Flora and fauna[1,2]	Habitat type; rarity; diversity; location of designated sites; level of designation
Birds[2]	Migratory birds; overwintering birds; protected species; level of designation
Fisheries[2]	Shellfisheries; commercial fish farms; fish nursery areas
Air quality	Ambient situation[3]
Water quality[1]	Ambient situation; trends; sources of pollution. Data in accordance with relevant EC Directive(s)
Sediment quality[1]	Level of contamination; historic records; land uses; pollution sources
Soil/ground conditions	Type; structure; bearing capacity; chemical composition
Geological interest[2]	Protected sites; exposures
Climate	Temperature; wind strength and direction; unusual characteristics
Hydrology/geomorphology	Drainage; surface runoff; channel characteristics; hydrological regime; sediment budget
Groundwater	Quality; status (e.g. aquifer)
Hydraulic regime[1]	Winds; waves; currents; tides; surges
Coastal geomorphology	Current status; processes shaping coast e.g. sediment transport; erosion rates; landslips
Population	Density; numbers; demographic features
Local community	Cultural status; interest groups
Housing	Existing stock locally; demand
Local economy	Economic well-being; major inputs
Employment	Major employers (dependence); unemployment rate; trends
Health and safety	Site records; special characteristics
Emergency services	Site records
Risk[1]	Existing sources of risk; characteristics; trends (e.g. standard of existing defence; current erosion rates; risk of incident)
Land use[1]	Urban; agriculture; military; infrastructure
Landscape[1,2]	Designated areas; topography; aesthetics
Leisure and amenity[2]	Seasonal use of site; footpaths; sailing and watersports; wildfowling
Tourism[1]	Tourist attractions; number of visitors
Heritage[2]	Archaeology; architecture; shipwrecks
Infrastructure	Roads; power; water
Traffic[1]	Road; rail; congestion; black spots; status in relation to capacity
Navigation	Main channels; history of incidents; status in relation to capacity; recreational and commercial use

Notes:
[1] Survey/assessment may be necessary
[2] Important sites generally already well documented
[3] Checklist is not intended to be definitive.

Overall, although the situation in respect of consultation is unclear, it should nevertheless be noted that a widespread and thorough consultation exercise can be extremely valuable throughout the EA process. Table 4.7 therefore provides a list of the primary agencies who might be among those consulted in respect of each of the environmental parameters listed. In addition, however, consideration should be given to consulting local groups and individuals (e.g. local wildlife trusts, birdwatchers, sailing clubs, anglers), and also consultants or contractors who have previous experience of working in the area.

4.3.6 Supplementary data collection

In addition to the collation of existing data (including that acquired via the consultation process) it should usually be anticipated that a certain amount of field survey and sampling will be required. This may be particularly important in respect of the following parameters where good quality, up-to-date information is often scarce.

- Water and sediment quality: a knowledge of the site and its environs (e.g. present and previous uses) together with the target EQSs or EQOs from appropriate EC Directives or similar documents should provide an indication of possible sampling and analytical requirements.

- Ambient noise levels: it is important to determine the effects of construction works (piling), traffic, etc.

- Ecological characteristics: marine ecological data, in particular, are rarely readily available from statutory/voluntary bodies.

- Hydrology, hydraulic regime, geomorphology: adequate and appropriate data on hydrological and/or coastal processes are not always available.

- Land use: may require mapping to establish, for example, the nature of assets at risk from flooding or erosion.

- Leisure, amenity and tourism: recreation and amenity use of a resource is usually poorly defined or recorded (e.g. number of people using a particular beach or footpath).

- Traffic: counts may be necessary to establish whether anticipated increases are likely to be significant.

- Archaeology: investigations may be necessary to establish the presence/absence of archaeological interest.

If the required information is not available, the developer should be aware that some data collection might take a considerable amount of time. A 'point-in-time' water quality sample or traffic count, for example, may not be appropriate, and a monitoring programme may be required to run for several months. Ecological data may need to be collected over a minimum of 12 months to ensure that breeding, feeding, roosting/nesting, migration, etc. are all covered and that the seasonal regime is properly established. Data on coastal processes will need to be collected over a number of months, usually including winter storms if possible. Conversely, any survey of tourism would ideally be carried out during the summer months. It is therefore essential that data collection is started at the earliest possible stage.

Table 4.7 Possible consultees in the EA process

Environmental parameter	Primary interested agencies
Flora and fauna/birds	English Nature, Countryside Council for Wales, Royal Society for the Protection of Birds, Royal Society for Nature Conservation, Wildfowl and Wetlands Trust, Woodland Trust, British Association for Shooting and Conservation, National Trust, Marine Conservation Society, local wildlife trusts
Fisheries	Ministry of Agriculture, Fisheries and Food, Sea Fisheries Districts, National Rivers Authority, Sea Fish Industry Authority
Air quality	Her Majesty's Inspectorate of Pollution, local authority
Water quality	National Rivers Authority, local authority
Sediment quality	Ministry of Agriculture, Fisheries and Food, London Waste Regulation Authority
Soil	Ministry of Agriculture, Fisheries and Food
Geological interest	English Nature, Countryside Council for Wales
Climate	Meteorological Office
Hydrology/geomorphology	National Rivers Authority, internal drainage boards, English Nature, Countryside Council for Wales
Hydraulic regime/coastal geomorphology	National Rivers Authority, local authority, Ministry of Agriculture, Fisheries and Food, English Nature, Countryside Council for Wales, private owners
Population/local community/ housing/local economy/employment	Local authority, Department of Employment, Chamber of Commerce
Health and safety/risk	Health and Safety Executive, local authority (environmental health officer)
Emergency services	Emergency services
Land use	Ministry of Agriculture, Fisheries and Food, local authority, National Farmers Union, Country Landowners' Association, Crown Estate Commissioners, CPRE, National Park Authority
Landscape	Countryside Commission, Countryside Council for Wales, local authority
Leisure and amenity	Local authority, Sports Council, Ramblers Association, Royal Yachting Association
Tourism	Local authority, Tourist Board, Sports Council
Heritage	Historic Buildings and Monuments Commission (English Heritage), local authority (county archaeologist), National Trust, Landmark Trust, Civic Trust
Infrastructure	British Telecom, British Gas, electricity Plcs, National Grid, water company, local authority, British Pipelines Agency, Ministry of Defence
Traffic	Local authority, Department of Transport, British Rail
Navigation	Port Authority, Harbour Commissioners, National Rivers Authority, British Waterways, Trinity House

4.4 IMPACT IDENTIFICATION, EVALUATION AND MITIGATION

4.4.1 Impact assessment process

In this section, the various subsections deal in broad terms with the identification and prediction of possible impacts. Ways in which the significance of such impacts might be assessed are explored, and if environmental quality standards or objectives exist, these are highlighted. Sensitive resources, which might be protected by some form of designation, are also discussed. Finally, potential mitigating measures for alleviating adverse impacts are suggested, and environmental enhancement options are presented.

The objective of an EA is to review all possible options and, unless none of the options is environmentally acceptable, select the option that causes the minimum (adverse) environmental impact. Consideration should therefore be given not only to the need for the proposed development but also, wherever this is practicable, to alternative locations and/or methods of achieving the desired aim. Each option should then be investigated in terms of its potential impact on the various characteristics of the existing environment.

Sections 4.4.2−4.4.12 deal with physical and chemical, biological and human environmental characteristics as defined in Box 4.1. In each case, where impacts are limited in either frequency of occurrence or potential scope, the issues involved are discussed briefly in the text. Where there are a large number of potential impacts associated with a particular parameter, examples of typical impacts are presented in a table. The following subsections and their associated tables are not, however, intended to be exhaustive. It is essential that, for each option under consideration as part of any proposed development, those undertaking the EA go through the process of identifying (using, for example, a matrix similar to Figure 4.2) and evaluating possible impacts, both on and off site. This process, as indicated in the various tables, should include:

- issue identification
- determination of cause of impact
- review of anticipated effects
- use of predictive techniques
- assessment of significance
- accordance with appropriate standards (if any)
- identification of mitigation and enhancement measures.

No direct reference is made in Tables 4.8−4.15 to whether a potential impact is direct or indirect and whether it is beneficial or adverse as these factors, particularly the latter, will be very much site specific. The loss of habitat, for example, could be beneficial if that which replaces it adds to the site's diversity, and the creation of jobs may not be beneficial to a local community if skilled workers have to be brought in from outside the area. The impact evaluation process should also consider whether impacts are long or short term, reversible or irreversible, and whether effects are local or widespread.

Finally, Sections 4.4.2−4.4.12 set out the process whereby impacts are identified and evaluated in terms of their significance. In many cases where potentially adverse impacts are identified, it should be anticipated that mitigating measures will be required. In addition to considering mitigation or ameliorative treatments, the EA process should also seek to identify and promote the implementation of environmental enhancement measures. Habitat creation, the provision of footpaths or seating, landscaping and the placing of information boards all represent opportunities for significantly enhancing the local environment, often at a relatively low cost.

Box 4.1 Categorisation of impacts arising from river and coastal engineering works

Main category		Section no.	Table no.
Physical and chemical effects	Water quality, water resources and groundwater	4.4.2	4.8
	Sediment quality, soil and geology	4.4.3	4.9
	Hydrology, geomorphology, hydraulic regime and coastal configuration	4.4.4	4.10
	Air quality	4.4.5	
Natural habitats	Flora, fauna, fish and birds Geology/geomorphology	4.4.6	4.11
Human characteristics	Socio-economic and local community	4.4.7	4.12
	Leisure and amenity/tourism	4.4.8	4.13
	Land use and landscape	4.4.9	4.14
	Health and safety; Emergency services; risk	4.4.10	4.15
	Traffic, transport, navigation and infrastructure	4.4.11	(4.13)
	Heritage and archaeology	4.4.12	

4.4.2 Water quality, water resources and groundwater

Water quality problems frequently arise in enclosed or semi-enclosed water bodies, particularly if flushing rates are low. Ports, harbours and marinas with restricted tidal access might therefore be expected to suffer from water quality problems as a result of either their own operations or those of others. Some ports and harbours, for example, may experience problems as a result of pollution entering the river further upstream (e.g. fertilisers from agricultural catchments). The construction of breakwaters for coastal defence or the construction of river control structures might similarly lead to the local or more extensive trapping of sewage and other pollutants.

Although river or coastal engineering works are unlikely to represent a source of pollution in themselves, they might indirectly introduce a potential source of pollution (e.g. a new oil tanker berth), or structures such as sluices or barrages may exacerbate an existing problem, for example by limiting water exchange. Table 4.8 outlines a number of surface and groundwater quality issues that might result from river and coastal engineering works, along with appropriate standards or controls and examples of enhancement or mitigation measures. It should be noted, however, that there are many other ways in which engineering schemes might directly or indirectly affect surface or groundwater resources.

Table 4.8 Examples of the impact of river and coastal engineering works on water quality and water resources

Issue	Possible cause	Typical effects	Predictive techniques	Appropriate standards	Mitigation and enhancement options
Acute chemical water pollution	Incident (e.g. collision; explosion; spill)	Poisoning of wildlife; potentially toxic substances introduced into food chain; contamination of water supply; contamination of sediments	Computer modelling (i.e. spill behaviour); hydrographic and oceanographic surveys	EC (Shellfish; Bathing Water; Drinking Water) Directives; NRA Water Quality objectives	Back-up safety systems; personnel education and training; effective (reactive) contingency plan; availability of booms, dispersants, etc.
Chronic chemical water pollution	Contaminated run-off; careless cargo handling; uncontrolled dumping of waste or disposal of bilge waters; leaching; enclosure of water body	As above; general breakdown of ecosystem	Long-term monitoring; hydrographic and oceanographic surveys	EC (Shellfish, Bathing Water, Drinking Water) Directives: MARPOL; UNCLOS (see Table 4.3); NRA Water Quality objectives	Personnel education; adequate flushing regime; on-site waste reception facilities; drip trays; effective interception/collection system; clean up existing problems
Eutrophication	Agricultural catchment; sewage discharge; problem exacerbated by barrage structure, tidal sill or other coastal structures	Algal blooms; death of aquatic flora and fauna	Monitor nitrate and phosphate levels	NRA Water Quality objectives	Improve flushing characteristics; reduce nutrient inputs at source
Turbidity	Dredging (capital or maintenance)	Smothering habitats; reduce feeding time for filter-feeding organisms	Monitor suspended sediment; hydrographic and oceanographic measurements	EC (Shellfish; Bathing Water, Drinking Water) Directives; NRA Water Quality objectives	Careful selection and control of dredging techniques; dredging on ebb tide
Resuspended sediments	Dredging; working in shallow water; propeller wash; bow thrusters	Release contaminants into water column for redistribution; poison aquatic life; potential toxins into food chain	Modelling; monitoring	NRA Water Quality objectives	Careful selection and control of dredging technique; treatment or removal of contaminants; armour seabed at berths; minimise turbidity near shell fisheries
Visual amenity	Visible surface pollution (e.g. oil); litter	Reduced landscape quality; aesthetic deterioration	Baseline assessment	Site specific	Clean up; ensure adequate screening; improve water circulation
Groundwater contamination	Leaking storage tanks; saltwater flooding or saline intrusion affecting aquifer; accidental spills	Contaminated drinking water supply; groundwater not suitable for irrigation purposes	Groundwater modelling	EC Groundwater and Drinking Water and Directives; NRA/MAFF irrigation water standards; discharges to aquatic environment	Good site management for prevention; flood protection for aquifer; reactive spill clean-up programme; adequate bunding of tanks
Seepage	Inappropriate foundations or piling cuts off or interrupts flow lines	Changed water quality characteristics immediately behind flood bank; loss of brackish water habitats	Habitat survey; soils investigation; modelling	Site specific	Maintain flow conditions/seepage if critical to habitat
Water demand	Temporary demand during construction; long term demand due to marina or port operations	Pressure on existing supply; requirements for new supply	Long/short-term demand analysis	Site specific	

4.4.3 Sediment quality, soil and geology

The primary impacts of river and coastal engineering works on sediments, soil and geological interest arise from either chemical contamination or physical damage. The prevention of the former can usually be achieved by good site management during both the construction and operational phases of a development. Once soil or sediment is contaminated, however, finding a means of disposing of the affected materials without causing either further contamination of the working site or undesirable consequences for the disposal site becomes critical. If a site is being dredged, there are further possible adverse impacts associated with turbidity and the release of contaminants into the water column. Various soil or sediment quality issues are considered in Table 4.9.

In addition to the potential impacts shown in Table 4.9, river and coastal engineering works could potentially have impacts on soil structure or slope (and hence structural) stability, and in some cases works might destroy rather than cover sites of geological interest. Where geological sites are protected by designation, detailed discussions with English Nature (Countryside Council for Wales) will be required in order to define a mutually acceptable way forward for the development.

Table 4.9 Examples of the impact of river and coastal engineering works on soil, sediments and geological interest

Issue	Possible cause	Typical effects	Predictive techniques	Appropriate standards	Mitigation and enhancement options
Contamination of soil	Leaking storage tanks or drums; uncontrolled dumping of wastes; spills	Renders soil potentially toxic (e.g. to vegetation); chemically unstable; health risk	Monitoring; use of vegetation as biological indicator	ICRCL Standards according to proposed land use	Personnel education; post-spill clean-up; waste disposal programme and adequate reception facilities; good site management
Disposal of contaminated soil/sediment	Excavation; cut and fill; imported materials; incident on site; dredging	Health risk; bio-availability of contaminants; possible consequences for ecology at disposal site	Sampling to determine type and degree of contamination prior to dredging/excavation	No British guidelines on sediment quality; ICRCL Guidance on the assessment and redevelopment of contaminated land	Treatment of soil; education to ensure careful handling and reduce health risks; select minimum-impact disposal site
Disposal of clean soil/dredged arisings	Excavation; capital or maintenance dredging	Smothering of habitat at disposal site; availability of materials for sea defence/coastal protection schemes	As above	As above	Habitat creation; coastal defence; other beneficial uses of dredged material
Structural stability	Exceeding bearing capacity; dredging; saturation; vibration	Structural cracking or failure of sea defence or port structures; land slip on cliffs	Geotechnical modelling	BS 8004; BS 8002 (draft)	Careful predictive studies; monitoring during construction
Soil structure	Compaction	Vegetation growth affected; unsuitable for agricultural use			Prevention; use-defined haulage routes; use of geotextile mattress; low ground pressure plant. Remedial measures: mechanical re-working; soil treatment
Covering of geological exposure	Beach recharge; construction of coast protection structure; land claim; cut and fill	No access to site; education/research value lost; prevents loss of site by erosion	Site inspection; discussions with county archaeologist/English Nature	Geological SSSIs or Conservation Review sites (GCR) indicate geological importance	Design scheme to prevent long-term damage; reduce rather than prevent erosion; careful instructions to contractor; expose face at alternative site; selective exposure *in situ*
Destruction of geological site	Excavation; cut and fill	Education/research value lost	As above	Geological conservation review sites and Geological SSSIs, indicate geological importance	Design scheme to prevent long-term damage; careful instructions to contractor; expose face at alternative site; relocate works

4.4.4 Physical processes: hydrology, geomorphology, hydraulic regime, coastal configuration

By their very nature, many types of river and coastal engineering schemes have the potential to interrupt or to change the natural physical processes operating at the site. Indeed, some schemes are specifically designed to initiate such a change. Of particular importance are the knock-on consequences of such works downstream, in terms of erosion and deposition characteristics, sediment supply, and the stability of slopes and structures. Impacts of river and coastal engineering works are considered in Table 4.10.

Box 4.2 Case study: Hydrological mitigation

The above works, completed by the National Rivers Authority Thames Region in 1991, form part of the Lower Colne Flood Alleviation Scheme. The scheme was designed to reduce flooding and provide a 1:100 year level of protection. The works involved widening the channel to accommodate the flood flow, but also incorporated an island to enable the retention of trees and create wildlife refuges. An Environmental Statement was required in this instance because of the significant impact associated with channel widening; however, this was mitigated by incorporating a low-flow channel within the widened section.

Courtesy: Dr Andrew Brookes, NRA - Thames Region.

Box 4.3 Case study: Physical process impacts

A proposed development within a SSSI on a Scottish estuary had it gone ahead as initially conceived, would have destroyed a habitat of national importance for overwintering Eider ducks. By relocating the land claim involved away from the mussel beds on which the birds were feeding, permanent damage was avoided.

Concern was still expressed, however, that fines from the reclamation process would settle out on the rocky outcrop, smothering mussels and hence damaging the birds' food source. The use of weir boxes was incorporated into the construction plan to minimise outwash of fines. Careful analysis of the physical processes operating locally demonstrated that the much reduced quantity of sediment that passed through the weir boxes would settle out before reaching the mussel beds. The Nature Conservancy Council then agreed to allow the project to proceed because conservation interests remained undamaged.

In addition to the typical impacts presented in Table 4.10, river and coastal engineering works might, directly or indirectly, cause several other impacts. Inter-tidal areas might be lost as a result of land claim, smothering and raising the site, or erosion due to dredging or the cutting off of the sediment supply. If salt marsh is lost, its buffering effect in terms of absorbing wave energy is reduced, and the risk of erosion might be expected to increase. A change in tidal volume resulting from dredging of siltation, or an artificial structure might also be expected to lead to a change in erosion/deposition patterns and, in some cases, to increased

Table 4.10 Examples of the impact of river and coastal engineering works on physical processes

Issue	Possible cause	Typical effects	Predictive techniques	Appropriate standards/ controls	Mitigation and enhancement options
Loss of inter-tidal area	Land claim; disposal of dredged arisings; dredging activities; erosion	Loss of buffer absorbing wave energy; increased erosion	Coastal process study; computer modelling	CIRIA *Seawall design guidelines*	Soft defences to absorb wave energy; environmentally sensitive engineering design; retreat to new defence line
Shingle disturbance	Shingle redistribution; beach recharge; plant movement	Loss of sensitive vegetation; loss of physiographic interest	Computer modelling; ecological survey		Careful planning; education of contractors on site; good site management
Change in channel morphology	New flood-alleviation channel; new port structure or land claim necessitating channel diversion	Downstream erosion; change pattern of deposition; change of velocity; change habitat type	Hydrographic measurements; physical modelling; computer modelling; ecological modelling; geomorphological analysis		Landscaping; habitat creation; flow control structures to prevent scour; recreational opportunities
Erosion	Dredging; boatwash; loss of sediment supply; natural state	Undercutting and collapse of embankments; loss of land and structures; loss of mudflats; steepening of beaches	Geomorphological analysis; computer or physical modelling; geotechnical analysis;	CIRIA *Seawall design guidelines*	Soft defences; vegetation planting; habitat creation and stabilisation; reorientation or removal of structures
Change in tidal volume	Dredging; siltation; new structure	Changing erosion/deposition; increased scour; ecological disturbance	Computer modelling		Habitat creation
Retained water levels	Barrage structure; tidal sill	Loss of inter-tidal area; reduced feeding opportunities for birds; disruption to fish migration routes and spawning grounds; sedimentation behind structure; recreation opportunities	Computer modelling; physical modelling; ecological survey		Creation of inter-tidal habitats as mitigation; minimise scale of loss of inter-tidal area; adequate flushing to prevent pollution; sensitive recreational use; dredging
Sediment supply	Groynes; breakwaters; beach recharge	Change in erosion/deposition characteristics; smothering of vegetation; creation of new inter-tidal or land areas	Computer modelling; physical modelling	CIRIA *Seawall design guidelines*; *Shore Protection Manual*; CIRIA *Use of Groynes*	Careful review of upstream/downstream implications of works; amend design to ensure continued sediment supply downstream if appropriate; by-passing
Instability of structures	Dredging; erosion; wave action	Failure of port structures, flood defence structures, coastal defence structures due to undercutting or washing out of fines from behind structure	Geotechnical/structural analysis	BS 6349; CIRIA *Seawall design guidelines*; CIRIA Publication 83	Improved structures; scour protection
Stabilisation of adjacent coastline; prevention of downstream erosion of land or flood defences	Coast protection or flood defence works	Coastal/riverside land uses protected; reduced stress and worry; change geomorphological characteristics	Physical modelling; computer modelling	Coast Protection Act 1949; Land Drainage Act 1991; CIRIA *Seawall design guidelines*; CUR Beach Nourishment; CUR/CIRIA *Use of Rock*; CIRIA *Use of Groynes*; BS 6349; *Shore Protection Manual*	Visually acceptable structures; minimise detrimental ecological/physical consequences

scour. Stabilisation of the coastline, while reducing the loss of land or the risk of flooding, might have similar downstream consequences. Computer modelling or physical modelling do offer opportunities to examine the effect of river and coastal engineering works on physical processes but it should be recognised that, if reliable results are to be obtained, data requirements are high and the possible cost implications of such modelling exercises should not, therefore, be overlooked.

4.4.5 Air quality

River and coastal engineering works rarely lead to significant changes in air quality, partly (for ports) as a result of existing controls. During the construction phase of works there may be some local deterioration in air quality due either to exhaust emissions (for example if large quantities of materials are being transported by road) or to dust resulting from site works. During the operational phase of port and harbour developments, there may be local impacts associated both with increased traffic to and from the port and with the handling of some bulk cargoes. Technical solutions such as water sprays (coal) or the use of air impellers and filters to create a negative air pressure around transfer points or points of discharge (grain) should, however, be available to mitigate the health implications of the latter. Road traffic problems can be countered in some circumstances by promoting the use of rail or sea transport.

If it is felt that there may be adverse knock-on impacts of a deterioration in air quality on the local community or on wildlife, however, a pre-development air quality survey should be undertaken. This is needed in order to establish ambient conditions, and post-development monitoring may be necessary to ensure that any predefined thresholds are not exceeded.

4.4.6 Natural habitats

In terms of nature conservation interests, the water's edge environment supports a wide range of sensitive and threatened habitat types. There are many ways in which river and coastal engineering works might impact directly (e.g. physical loss) or indirectly (e.g. pollution affecting the source of food) upon the flora and fauna of our coasts and rivers. Damage to or loss of bird habitats is most commonly cited, but the aquatic or marine resource can also be affected by works. Relatively little is known about the marine environment and the assessment of potential impacts can, therefore, be relatively difficult and/or time consuming. Fisheries interests can also be affected by river and coastal engineering works. Shellfish and other filter feeders might be affected by turbidity. Water pollution can lead to contaminants entering the food chain, ultimately affecting both wildlife and humans. Table 4.11 presents a range of the typical impacts, both direct and indirect, of river and coastal engineering works on flora and fauna, fish and birds.

Table 4.11 Examples of the impact of river and coastal engineering works on natural habitats

Issue	Possible cause	Typical effects	Predictive techniques	Appropriate standards/ controls	Mitigation and enhancement options
Removal of communities	Dredging; channel clearance; maintenance of structures	Loss of river or seabed habitat; disturbance of species attached to quay walls etc.	Site survey	Sensitivity of site and species indicated by site survey and/or official designation	Habitat creation including reefs, fish habitat, etc; phased works to enable recolonisation
Turbidity	Dredging	Reduced light penetration leading to a reduction in photosynthesis; feeding and respiration difficulties for fish; smothering of benthos	Computer modelling; ecological monitoring		Plan dredging activities to avoid sensitive times of year (e.g. overwintering; breeding season, etc.)
Loss of inter-tidal area (mud-flats, salt-marshes etc.)	Dredging; land claim	Loss of mud-flats and salt-marshes through erosion or smothering; reduce food availability for birds	Geotechnical slope stability analysis and/or modelling to predict extent of loss	Many British estuaries are now designated SSSIs	Eliminate unnecessary dredging; habitat creation as mitigation
Shingle habitat disruption	Shingle redistribution; removal for recharge scheme; plant movement	Loss of shingle vegetation and geomorphological features	Regular monitoring	SSSI protection afforded to many shingle features	Careful planning; education of contractors on site
Loss of bankside/ terrestrial habitats	Lowering of water-table; site preparation; surfacing; materials extraction from borrow pits	Loss of wetland communities and other wildlife interest; loss of wildlife corridor along riverbank; loss of aquatic communities	Site survey	Water Resources Act 1991	Minimise loss; work from one bank only; habitat creation or restoration; phased works to facilitate recolonisation
Disturbance	Construction process (e.g. piling); noise; lighting	Disrupts bird feeding, breeding, and roosting patterns; disturbs wildlife	Monitoring before and during construction		Minimise night-time working; careful timing of works; screen works if possible; angle lights away from sensitive sites
Physical damage	Careless mooring; unauthorised access routes; unnecessary vegetation removal; replacing natural vegetated slope with hard or vertical defences	Disturbance; habitat loss; loss of wildlife corridor	Ecological survey	S16, Water Resources Act 1991	Replanting; creation of new habitats
Vegetation damage	Traffic; dust; exhaust fumes; weed cutting	Damage to terrestrial vegetation; disturbance to fauna; removal river bed/bank vegetation	Site survey	S16, Water Resources Act 1991	Improve fuel efficiency; use rail or water transport; work from one bank only; leave some aquatic habitats intact
New or improved habitats	Structures; earth embankments; soke dyke creation; rock armour; borrow areas	New surface for aquatic organisms; new or extended wetland habitats	Site survey; topographic survey	Water Resources Act 1991	Careful selection of appropriate seed mix; tree species; landscaping; planting of borrow pits, etc.
Inundation	Failure of flood defence structure	Habitat change/loss; increased salinity	Ecological modelling		Encourage beneficial inundation; habitat creation
Water pollution (see Table 4.8)					

There are many potential consequences of such works on natural habitats. Physical damage to habitats might include the removal of aquatic communities by dredging, maintenance of structures including quay walls, or weed-cutting; the loss of wetland habitats due to the disposal of dredged material, dumping of rubbish, or the construction of sea defences; and the loss of terrestrial habitats through vegetation stripping or the digging of borrow pits. Many of these activities, however, may create enhanced replacement habitats (for example, ponds, lagoons, or bird scrapes) and whether the overall impact of works is beneficial or adverse will, therefore, depend on the original status of the site in terms of habitat diversity, rarity, etc. Some habitats are particularly sensitive to disturbance. Shingle vegetation, for instance, takes a long time to become established and stable. A bulldozer can very quickly destroy a shingle community.

The loss of a wildlife corridor along a river bank by installing a length of steel sheet piling is similarly very difficult to restore but, in certain special cases, if piling is the only viable engineering option to prevent a valuable freshwater habitat such as that found in Norfolk's Broadland from being inundated with saltwater, the loss might be judged to be acceptable by those responsible for nature conservation. Detailed consultation with both statutory and voluntary nature conservation bodies should be regarded as an integral part of most river and coastal engineering schemes.

Finally, the option of habitat creation or restoration is mentioned several times in Table 4.11. It should be noted, however, that while the creation of some types of coastal and riverside habitats (e.g. reedbeds, sand dunes, scrapes) is relatively well researched, little is known about the long-term ecological viability of created salt-marshes or mud-flats. The conservation of existing habitats of this type should, therefore, be regarded as a high priority when coastal engineering works are being planned.

Box 4.4 Case study: Habitat protection

At Harwich Parkeston Quay, English Nature were concerned that the capital dredging required as part of the construction of a new oil tanker berth would lead to the loss of a significant area of mud-flats in the Stour Estuary Site of Special Scientific Interest. A detailed geotechnical study demonstrated that the total inter-tidal area lost would in fact be minimal, but the Environmental Statement also identified a need for the development of an oil spill contingency plan. This was considered essential to ensure that possible damage to the internationally important wildlife resource would be minimised in the event of an oil spill.

4.4.7 Socio-economic and local community

Flood defence and coast protection schemes will usually have only a limited impact on socio-economic parameters. Any employment that is created as a result of the construction process is likely to be relatively short term and while there may be some knock-on economic benefits locally, these are not likely to be of particular significance. Port, harbour and marina developments, however, can sometimes lead to significant social changes. The EA process should therefore include a thorough review of the primary, secondary and tertiary socio-economic implications of any development. The creation of jobs may initially appear beneficial, but demographic and/or employment modelling may reveal undesirable knock-on consequences in the form of pressure on schools and health services, or a shift in the economic or cultural base of a community. Similarly, the creation of jobs may be at the expense of a loss of wildlife resources. In this case all possible mitigation and enhancement options must be thoroughly explored and implemented. Table 4.12 demonstrates some typical socio-economic impacts that might result from river and coastal engineering works.

Table 4.12 Examples of the impact of river and coastal engineering works on socio-economic parameters

Issue	Possible cause	Typical effects	Predictive techniques	Appropriate standards/ controls	Mitigation and enhancement options
Disturbance	Noise (e.g. piling); light; vibration; traffic	Stress; worry; loss of sleep; safety concerns	Background measurement; experience at similar sites	BS 7385 BS 7482 BS 5228 (Parts 1, 2, 4)	Daytime working; careful selection of piling techniques; screening; angled of lighting; double glazing; routeing of traffic to avoid sensitive areas
Employment opportunities	Construction work; port or harbour expansion; marina development; support industries	Local job creation; immigration; boost to local economy but large construction project may create non-sustainable 'boomtown' economy	Economic modelling; employment forecasting	Case specific	Provide skills training for local people
Unemployment	Port rationalisation/ automation; end of major construction project	Social problems; demoralisation, dereliction, vandalism; economic stagnation; emigration	Socio-economic modelling	Case specific	Local or central government employment initiatives
Immigration	Employment opportunities	Pressure on existing short/long-term accommodation, schools, health services, etc.	Demographic modelling	Case specific	Contractor to provide temporary accommodation; encourage use of local labour; provide skills training; developer to fund/contribute to new services etc.
Increased property prices/rents	Increased demand; change in status of site due to new marina or redevelopment of port/ harbour	Local people or business priced out of market	Supply/demand analysis	Case specific	Provide concessions for local businesses
Changed cultural/ economic base	Major development or redevelopment (e.g. new port; marina; etc.)	Change social class through immigration; rural to urban society; social upheaval; attract tourists/visitors to area	Demographic modelling; economic analysis	Case specific	Retain local cultural features; respect local customs; establish and provide for local needs
Health benefits	Reduced flooding/erosion risk	Reduced stress and worry		Case specific	Good dissemination of information to public; effective emergency plan; trained personnel

In addition to the impacts set out above, river and coastal engineering schemes can potentially cause disturbance to the local community as a result of traffic, noise, vibration and/or dust. With most flood defence and coast protection schemes, such impacts are generally short term as a result of the construction process. With port and harbour engineering, however, traffic-associated disturbance may be one of the major long-term impacts of a development.

4.4.8 Leisure and amenity/tourism

In the UK a great deal of both formal and informal recreational time is spent on or adjacent to water. Water-based recreational activities such as sailing, windsurfing and canoeing are increasing in popularity; river bank and coastal footpaths are heavily used; and the demand for clean, sandy beaches is consistently high.

Table 4.13 Examples of the impact of river and coastal engineering works on leisure and amenity

Issue	Possible cause	Typical effects	Predictive techniques	Appropriate standards/ controls	Mitigation and enhancement options
Access	Site works; extension of area of permanent operations	Temporary or permanent closure of footpath or vehicular access; requirement for diversion; reduced opportunity for informal recreation; loss of public access to development site	Modelling pattern and intensity of use	Temporary or permanent diversion order from Highway Authority; CIRIA *Seawall design guidelines*	Well-signposted, well-maintained diversion; new or reinstated footpath accommodating disabled users; new link road/bypass if appropriate
Construction traffic	Movement of materials to site	Traffic congestion; damage to roads; exhaust emissions; vibration; noise	Traffic modelling	Agreement with County Highways and other authorities on routeing, etc; CIRIA *Seawall design guidelines*	Careful programming of works; ensure all materials are on site before start of tourist season; careful choice of construction traffic route; reinstate road
Land requirements for contractor	Stockpiling; site works; offices; car parking	Temporary loss of car parking/boat storage areas; temporary or permanent loss of informal recreation area; reduced tourism revenue	Site specific	S16 Water Resources Act 1991; CIRIA *Seawall design guidelines*	Consultation to ensure works take place outside peak season; provision of alternative parking area; cleaning up and reinstatement of site after use; planting and enhancement works
Change in beach area	Placing of rock; beach recharge	Loss or gain of beach area with implications for recreation/amenity use	Pattern/ intensity of use	S16 Water Resources Act 1991	Minimise area lost; ensure adequate beach area exposed at high tide
Visitor attractions	Site works; beach recharge; marina development	Tourists/visitors observing construction works (for safety implications see Table 4.15); beach enhancement; improved recreation resource; increased number of users; car parking/infrastructure requirements; services; boost to local economy	As above		Public display boards/exhibition; beach management plan; define acceptable uses; well-signposted facilities
New opportunities for water-based activities	Contained water body behind barrage or similar structure; new flood alleviation channel	Opportunities for sailing, windsurfing, canoeing	Supply/ demand analysis	Water Resources Act 1991	Landscaping; mitigation for habitat loss; provision of footpaths and slipways

It is inevitable, therefore, that river and coastal engineering works will impact upon recreation and tourism, particularly during the construction phase of works. The degree of disturbance caused can, however, be limited in most cases, provided that surveys have been carried out sufficiently early in the design process to establish the pattern of use of the particular resource (e.g. frequency, density and seasonality). Alternative points of access, to the riverside or beach for example, can be provided. Notice boards and announcements on local radio will help to keep the public informed of the progress of major works. Working outside the tourist season may be possible in many cases.

In terms of operational impacts, the use of beach nourishment as a coastal defence option could provide recreation and amenity benefits into the long term. Similarly, the provision and maintenance of footpaths, slipways, beaches or angling platforms might provide a significant enhancement of the amenity resource, in turn helping to mitigate any detrimental impact during the construction phase. Table 4.13 demonstrates some typical impacts of river and coastal engineering works on the recreation resource.

Box 4.5 Case study: Recreation benefits

An Environmental Statement produced as part of the development of a long-term strategy for sea defences along the Lincolnshire coast identified a number of potential impacts associated with both the construction and operational phases of coastal engineering works. The importance of minimising disruption during the construction process along this popular holiday coast was stressed by those consulted. The beach recharge option, however, attracted widespread support, representing an improved amenity and recreation resource in the longer term as well as being less visually intrusive than concrete sea walls. The Environmental Statement also identified already congested access routes that should be avoided, particularly during the holiday season, and recommended that as many materials as possible should be imported by sea rather than by road.

4.4.9 Land use and landscape

Most river and coastal engineering projects result in some impact upon land use and landscape, if only in the short term. With flood defence and coast protection schemes, impacts will range from a temporary change in land use, associated with site works for example, to a major impact upon the landscape such as might be caused by the construction of a flood defence barrier or barrage structure. Port and harbour expansion developments can also cause significant impacts, particularly where greenfield developments or major reclamation schemes are proposed. Although in some cases land-based port expansion might affect agricultural or amenity land, the issue that has caused most controversy over recent years is the loss of inter-tidal areas as a result of either land claim or dredging. The Royal Society for the Protection of Birds, for example, suggest that 90% of the inter-tidal area of the Tees Estuary has been lost as a result of land claim, mainly for industrial use, over the last 100 years.

The Countryside Commission (1987) published guidelines on the assessment of landscape impacts. These guidelines are intended for general use but, together with local planning authority controls (e.g. through structure plans/local plans/etc.), become particularly important in areas designated as being of special landscape significance. Countryside Commission publication CCP 326 is also relevant. In coastal and riverside areas, the following designations are especially relevant to landscape, among other planning issues:

- Heritage Coast
- National Parks
- Areas of Outstanding Natural Beauty
- Environmentally Sensitive Areas
- Special Landscape Areas.

Box 4.6 Case study: Landscaping opportunities

A public consultation exercise undertaken in order to ensure that a new marina development at Pwllheli in North Wales met local requirements, as well as catering for visitors, revealed a demand for areas of public open space and highlighted the need for the sympathetic landscaping of the development. As a result, a salt-tolerant seed mix was selected for the reclaimed land and areas were set aside as parkland. Recommendations were also made that the design of any new workshops or factory units should take account of the magnificent landscape setting of the development.

In most situations, however, landscaping measures will offer some scope for reducing adverse impacts or promoting environmental enhancement. Landscaping opportunities are therefore among the issues highlighted in Table 4.14, which outlines a selection of typical potential land-use and landscape impacts.

4.4.10 Health and safety; emergency services; risk

As with any civil engineering project, river and coastal engineering works will potentially impact upon health and safety and risk characteristics during both the construction and the operational phases of a project. A new facility (e.g. a port) or structure (e.g. flood defence) might change the likelihood of a particular occurrence. A new berth might increase or reduce the risk of collision, depending on its location and the number of vessel movements in the vicinity. Similarly, the risk of an incident such as a spillage or explosion might change. A risk assessment will indicate the probability of such an event, but a well-rehearsed contingency plan and good communications between vessels and the shore will be essential in most cases.

Waterborne plant (e.g. piling rig or barge importing rock armour) might also lead to a temporary increase in the risk of a collision with its attendant consequences. Traffic and spillage are dealt with in Tables 4.15 and 4.8 respectively. Finally, in terms of risk characteristics, improved sea defence or coast protection standards might significantly reduce the risk of flooding or the rate of erosion, usually with beneficial consequences.

In terms of health and safety requirements, it is assumed that both the construction and the operation of river and coastal engineering projects will follow all the appropriate HSE and BSI guidance. Accidents could potentially occur, however, as a result of machinery malfunction, structural or mechanical collapse, human error, plant movement, increased traffic or natural hazards, notably storms or floods. This potential for injury or loss of life, in addition to requiring trained personnel on site, puts pressure on the existing emergency services or introduces a need for new services. Good site management, education and training of all personnel, and implementation of all regulations should, however, ensure that most accidents can be avoided.

Table 4.14 Examples of the impact of river and coastal engineering works on land use and landscape

Issue	Possible cause	Typical effects	Predictive techniques	Appropriate standards/ controls	Mitigation and enhancement options
Construction site	Plant; stockpiling; structures; site offices	Visual impact; temporary loss of amenity/agricultural land; focus of interest for locals/visitors	Landscape assessments; Countryside Commission CCD18 (1987)	Local authority or HSE controls	Minimise area affected; screening; reinstate and improve site after construction
Permanent visual intrusion	Structures; operations	As above but long-term effect; 'blot on the landscape'	As above	As above; planning permission conditions	Screening; tree planting; careful selection vegetation; colouring; design aesthetics
Waste disposal	Disposal of excavated or dredged material, surplus construction materials, by-products; boat-generated wastes	Smother vegetation; loss of visual amenity; soil contamination; loss of previous land use; surface/groundwater contamination	Type and quantity of waste	MARPOL; UNCLOS (see Table 4.3); Oslo and London Dumping Conventions	Good site management; impose controls on illegal/irresponsible dumping; promote waste reception facilities on site
Permeability	Surfaces; slopes; structures	Increased surface runoff; loss habitat; change land use	Hydrological modelling		Vegetated slopes; porous surfaces; adequate drainage with interceptors
New land	Land claim	Loss of inter-tidal area; change coastal configuration; geomorphological impacts	Physical models; computer models; geotechnical studies	Designation of nature conservation site	Habitat creation
Land-use intensification	Improved flood defences; secondary development associated with new marina or port	Change low- to high-intensity agriculture; urban sprawl; demands on services	Behavioural studies	Planning permission/ controls	Incentive to farmers to maintain traditional systems; carrot and stick approach by local authorities to encourage development or not, as appropriate

Table 4.15 Examples of the impact of river and coastal engineering works on safety characteristics

Issue	Possible cause	Typical effects	Predictive techniques	Appropriate standards/ controls	Mitigation and enhancement options
Accidents affecting humans	Machinery malfunction; structural/mechanical collapse; human error; natural hazards; increased traffic; plant movement	Injury or loss of life; requirement for trained personnel on site; pressure on emergency services or need for new services	Risk assessment	HSE Site Safety guidelines; company site safety manual; BSI	Good site management; personnel education and training; implementation of regulations and guidelines
Water safety	New structures; deeper water following dredging; vertical walls	Attractiveness to children; increased risk of drowning	Risk assessment		Signing; lifesaving facilities; chains on concrete-lined walls; ladders and hooks on vertical walls
Changed risk characteristics	New facility; change in use of existing facility	Increased or reduced risk of collision, explosion, pollution etc.	Risk assessment		Contingency plan; good communications network
Spillage	See Table 4.8				

Finally, it should be acknowledged that some water's edge developments may increase the risk of accidents, notably involving children, and possibly loss of life through drowning. A new vertical quay wall should therefore have adequate handrailing and exit ladders. Lifebelts and other facilities should also be available nearby.

4.4.11 Traffic: transport, navigation and infrastructure

As mentioned earlier, river and coastal engineering schemes have the potential to disrupt access and generate traffic, particularly during the construction phase, but also during the operational phase of port, harbours and marina developments. Port developments in particular will need to incorporate a thorough assessment of the potential impacts of any increased traffic generation. The environmental impacts of highway projects are discussed in Chapter 6.

Although river and coastal engineering schemes can potentially have a number of adverse environmental impacts associated with traffic and transport, they can in some cases provide significant benefits. Bypasses and port link roads can help to reduce congestion; footpaths along defences or barrage structures can provide a new amenity or recreation resource, and some schemes may offer an opportunity to provide access to the waterside for the disabled. Access issues were discussed in Table 4.13.

4.4.12 Heritage and archaeology

River and coastal engineering works can potentially impact upon heritage and archaeological interests in a number of ways. Piling operations or traffic movements can cause vibration and structural damage or even collapse. Erosion, induced by dredging for example, could cause the undercutting or loss of features of interest. Both of these impacts should, however, be largely avoidable.

Most potentially damaging to heritage interests is the land requirement of many developments, notably new port or harbour facilities or marinas. This may result in the loss of a feature due to the space required to be developed, or because of the need for stability and hence foundations. If the site is already listed or designated in some way, a variety of controls exist to protect the feature(s) in question, but in most cases careful design should ensure that it (they) are protected. If previously unknown features of interest are found part way through the construction process, however, considerable delays might be expected. The County Archaeologist and various voluntary agencies (see Table 4.7) should be able to advise at an early stage whether or not there is (likely to be) significant heritage interest at a particular site.

4.5 POST-PROJECT ASSESSMENT REQUIREMENTS

Once the impact identification and evaluation process is complete, there are a number of other steps which must be taken before an ES is produced. First, it is essential that the residual impacts of each development option, particularly the preferred option, are clearly identified and that any impacts that depend on mitigation to reduce their severity are highlighted. This information, which should be presented clearly and impartially, is vital to the decision maker. Secondly, and of particular importance where there is uncertainty in respect of the impact prediction process, a monitoring process should be established that will provide information on such aspects as vegetation development and rates of colonisation, downstream sediment movement, turbidity during construction, or erosion rates. Alternatively or additionally it may involve on-site supervision of the construction process by an ecologist. Environmental assessment is a young discipline, and many of the predictive methodologies available should be treated with caution and assumed to be potentially imprecise.

Post-project appraisal or, if appropriate, the audit of a site or development, represents the third 'post-assessment' step. This involves not only an overall review of the development, but also the testing of predictions against the reality of their impact. Provision for, and possibly the programming of, such a review(s) should be carried out within the EA process to ensure that the need for such a review is acknowledged and documented.

Finally, a programme of future liaison and consultation must also be determined and set out in the ES. This is very important if the good relations developed during the preparation of the ES are to continue into the future. If local residents are kept informed about the programme and progress, they are less likely to be unduly concerned about a delay. Notices in newspapers and on notice boards, announcements on local radio, and public meetings all help to ensure the smooth implementation of the proposed scheme. Best practice for EA dictates that members of the public or their representatives are consulted at every stage of the development process. Good public relations are critical to good environmental management, both on a short-term (i.e. project) basis and in the longer term.

4.6 REFERENCES AND BIBLIOGRAPHY

Directives

Directive concerning the quality of bathing water
(76/160/EEC)

Directive concerning the quality required of surface water intended for the abstraction of drinking water
(75/440/EEC)

Directive on pollution caused by certain dangerous substances discharged into the aquatic environment of the community
(76/464/EEC)

Directive on the assessment of effects of certain public and private projects on the environment
(85/337/EEC)

Directive on the protection of groundwater against pollution caused by certain dangerous substances
(80/68/EEC)

Directive on the quality of fresh waters needing protection or improvement in order to support fish life
(78/923/EEC)

Directive on the quality required of shellfish waters
(79/923/EEC)

Habitats and Species Directive
(92/43/EEC)

Wild Birds Directive
(79/409/EEC)

Statutory Instruments

The Harbour Works (Assessment of Environmental Effects) Regulations 1988 No. 1336

The Harbour Works (Assessment of Environmental Effects) (No. 2) Regulations 1989 No. 424

The Land Drainage Improvement Works (Assessment of Environmental Effects) Regulations 1988 No. 1217

The Town and Country Planning (Assessment of Environmental Effects) Regulations 1988 No. 1199

Publications

CIRIA (1992a)
Seawall Design Guidelines
CIRIA (London)

CIRIA (1992b)
Manual on the Use of Rock in Coastal and Shoreline Engineering
CIRIA Special Publication 83

COPPIN, N.J. and RICHARDS, I.G. (1990)
Use of Vegetation in Civil Engineering
CIRIA (London)

COUNTRYSIDE COMMISSION (1987)
Landscape Assessment: A Countryside Commission Approach
CCD18

DENNY, P. (1993)
A Guide to Environmental Procedures for Inland Flood Defence
English Nature (in preparation), Peterborough

ENGLISH NATURE (1994)
Guidelines for Environmental Assessment
(in preparation)

GARDINER, J.L. (1991)
River Projects and Conservation: A Manual for Holistic Appraisal
Wiley and Sons

HAYNES, R. (ed) (1982)
Environmental Science Methods
Chapman & Hall

HMSO (1988)
Calculation of Road Traffic Noise

HMSO (1990)
Environmental Assessment: A Guide to the Procedures
HMSO

HYDRAULICS RESEARCH (1991)
Review of Novel Shore Protection Methods
HR Wallingford

ITE/NERC (1986)
Coast Dune Management Guide
ITE/NERC, Huntingdon

KING, A. and WATHERN, P.
Environmental Assessment of NRA Projects

LEE, N. and COLLET, R. (1990)
Reviewing the quality of environmental statements
University of Manchester Occasional paper 24

NCC (1984)
Nature Conservation in Britain
Nature Conservancy Council (English Nature), Peterborough

O'RIORDAN, T. and TURNER, R.K. (1983)
An Annotated Reader in Environmental Planning and Management
Pergamon Press

RSPB/RSNC (1984)
Rivers and Wildlife Handbook

WATHERN, P. (1988)
Environmental Assessment Theory and Practice
Unwin Hyman

Environmental statements

Continental ferry port phase VII, Portsmouth Harbour
Environmental statement, December 1992
Posford Duvivier Environment

Mablethorpe to Skegness sea defence strategy
Environmental statement, January 1992
Posford Duvivier Environment

Mablethorpe to Skegness sea defences. Offshore dredging licence application
Environmental statement, November 1992
Posford Duvivier Environment

Proposed oil berth development at Parkeston Quay, Harwich
Environmental statement, July 1991
Posford Duvivier Environment

Proposed ro-ro berth, passenger and cargo berth at Falmouth Docks, Falmouth
Environmental statement, October 1992
Posford Duvivier Environment

Engineering and British Standards

BS 5228 Noise control on construction and open sites
Part 1 Code of practice for basic information and procedures for noise control
Part 2 Guide to noise control legislation for construction and demolition including road construction and maintenance
Part 4 Code of practice for noise control applicable to piling operations

BS 6349 Maritime structures Part 5 Code of practice for dredging and land reclamation

BS 7385 Guide for measurement of vibrations and evaluation of their effects on buildings

BS 7482 Instrumentation for the measurement of vibration exposure of human beings

BS 8004 Foundations

CP2 Earth retaining structures (Draft BS 8002)

Papers

BROOKE, J.S. (1990)
Environmental appraisal for ports and harbours
The Dock and Harbour Authority

BROOKE, J.S. and PAIPAI, H. (1990)
Pollution control in ports, docks and harbours
The Dock and Harbour Authority

BROOKE, J.S. and WHITTLE, I.R. (1990)
The role of environmental assessment in design and construction for flood defence
Institution of Water and Environmental Management Conference, Glasgow, September 1990

BROOKES, A. (1991)
Environmental assessment in modern river management in the Thames region of the National Rivers Authority
International Symposium on Effects of Watercourse Improvements

CIRIA Research Project 424

Environmental Assessment

A guide to the identification and mitigation of environmental issues in construction schemes

5. Water supply infrastructure and wastewater treatment works

Carolyn Francis, Environmental Assessment Manager, Sir William Halcrow & Partners

5 Water supply infrastructure and wastewater treatment works

5.1 TYPE OF SCHEME CONSIDERED

This chapter outlines the procedures for the assessment of the environmental effects of constructing, operating and decommissioning water supply infrastructure projects and wastewater treatment works. The works involved in such schemes vary enormously depending, among other things, on the nature of the scheme, its size and the type of treatment. Table 5.1 summarises the most common structures that would be built for such schemes together with typical promoters.

Table 5.1 Examples of water supply infrastructure and wastewater treatment schemes

Scheme type	Typical promoter	Types of works	Type of structure/ solution
Water supply	Water company; industry (including power generators); NRA	Impoundments	Dams; spillways; draw off towers; pumping stations; hydro-electric station; pipelines; access road; amenity development; washlands
	Water company; farmers; fish farms; industry; NRA	River and groundwater abstractions	River intake; borehole; pumping stations; pipelines and river basin transfers; storage tanks or reservoirs; deepening ditches
Water and wastewater treatment	Water company; industry	Water treatment works	Treatment works; power supply; pumping station; storage tanks; public water supply; system (pipelines); access roads
	Water company; industry	Wastewater treatment works	Sewerage system; inlet works; screens; tanks for settlement; biological and chemical treatment; sludge storage; pumping station; power supply; river outfalls; vegetative treatment systems
	Water company; waste disposal companies	Sludge volume reduction	Buildings housing dryer/ incinerator; chimney stack; sludge storage and handling facilities
	Water company; industry	River and sea outfalls	Pumping station; power supply; outfall; diffuser heads

This chapter concentrates on the environmental assessment of reservoirs, water and wastewater treatment works, sludge treatment and outfalls. Storm overflows are not specifically covered. The effects of pipelines and industrial wastewater treatment works will be covered to a lesser extent, although the former are dealt with in Chapter 6. The environmental impacts of related subjects such as irrigation systems, river structures related to intakes for abstractions, agricultural and urban drainage, and reservoirs built for flood protection will not be considered. The types of environmental impact and mitigation method described relate to experience in the UK.

5.2 KEY STANDARDS AND LEGISLATION

5.2.1 Legislation governing Environmental Assessments

The European and UK legislation on environmental assessments (EA) is outlined in Section 2.1. Under the Council Directive (85/337/EEC) none of the types of works described in this chapter would seem likely to class as Annex I projects for which an environmental assessment is mandatory. However, Annex II projects for which an environmental assessment is discretionary include:

- dams and other installations designed to hold water or store it on a long-term basis
- installations for the disposal of industrial and domestic waste
- wastewater treatment plants
- sludge deposition sites.

In the UK, the EA of water supply infrastructure schemes and wastewater treatment works that require planning permission is implemented by the Town and Country Planning (Assessment of Environmental Effects) Regulations 1988 (SI No 1199), which covers England and Wales, the Environmental Assessment (Scotland) Regulations 1988 (SI No 1221), and Planning (Assessment of Environmental Effects) Regulations (Northern Ireland) 1989 (SR No 20). Annex I projects are defined in the regulations, but the decision as to whether a project falls under Annex II lies with the local planning authority. The developer may appeal to the Secretary of State for the Environment against a decision made by the local planning authority to undertake an EA (Section 2.1.4).

Requirements on the scope of the EA and the content of the environmental statement (ES) in which the results of the EA are reported follow those set out in the Council Directive as described in Section 2.2. The ES is submitted to the local planning authority together with the planning application. Copies are sent, either by the applicant or the local planning authority, to the statutory consultees, who have an opportunity to comment. The ES also has to be available for public inspection. The local planning authority takes into consideration the representations from the public and statutory consultees prior to giving a decision on the planning application. These procedures are described fully in the DoE Welsh Office publication *Environmental Assessment: A guide to the procedures*.

The Planning and Compensation Act 1991 affects the EA regulations in Sections 18 and 48 by making provision for the Secretary of State to add to the classes of projects requiring an EA. The Department of the Environment has since issued proposals for adding seven classes of projects including trout farms and water treatment works.

5.2.2 Legislation on conservation issues

The Wildlife and Countryside Act 1981 (as amended 1985) is the single most important piece of legislation on nature conservation in the UK (Table 5.2). It incorporates previous legislation, such as the Protection of Birds Acts 1954 and 1967, and the Conservation of Wild Creatures and Wild Plants Act 1975, as well as giving legal protection to species listed in the Wild Birds directive (Table 5.3) and Berne Convention (Table 5.4). The Act is divided into three parts.

Part I deals with the protection of birds, other wild creatures and plants. Animals are protected from being killed, injured or taken for sale, as are eggs and nests, and other structures or places used by animals. There are three exceptions to this: animals protected by other Acts of Parliament, the taking of injured animals to kill them humanely, and as the 'incidental result of a lawful operation' which could not have been reasonably avoided. An additional level of protection is afforded to rare species of plants and animals, which are listed in several schedules.

Part II covers the conservation of the countryside. It includes:

- the designation of Sites of Special Scientific Interest (SSSI) which were first designated under the National Parks and Access to the Countryside Act 1949

- the obligations on the Nature Conservancy Council (and its successors following the Environmental Protection Act 1990) to inform interested parties of notification

- procedures to be followed should the owners or occupiers wish to carry out a potentially damaging operation (PDO) as defined by the Nature Conservancy Council.

The DoE Circular *Nature Conservation* (27/87) states that local and county plans need to adopt policies to protect SSSIs, and, following the Town and Country Planning General Development Orders 1988, local planning authorities have to consult the Nature Conservancy Council over any planning applications affecting a SSSI. This will be superseded in 1994 with the publication of a PPG on Nature Conservation.

Part III of the Wildlife and Countryside Act 1981 deals with access to the countryside and rights of way.

There are separate UK Acts of Parliament that deal with the protection of specific animals, notably deer, seals and badgers. These would have to be considered in detail should a development be likely to affect these animals.

The Council Directive on habitats (92/43/EEC) is the latest initiative on conservation of wild fauna and flora and habitats. Annex I of the directive lists habitat types and Annex II lists species of flora and fauna that require the designation of special sites for their conservation. Member states are required to propose sites (including those already designated under the Wild Birds Directive 79/409/EEC) for conservation. Once a site has been adopted as being of community importance by the Commission, the member state has to adopt measures to designate and protect the site and species identified in Annex IV.

The conservation of marine environments around England and Wales has been considered further by the publication in February 1992 of a draft paper on Marine Consultation Areas by the Department of the Environment and Welsh Office. This aims to protect important communities of marine flora and fauna; breeding, wintering and feeding areas; and integrate conservation across the tidal margin where a SSSI adjoins the sea. It depends on the adoption of a voluntary consultation procedure between developers and either English Nature or the Countryside Council for Wales. Annex B identifies operations for which consultations should take place, including:

- proposals for a development on land that may have a significant impact on a marine consultation area and which is to be carried out through planning permission

- consents for effluent discharge including industrial and transport sources, sewage and cooling water, but excluding discharge from individual domestic properties and businesses of a comparable size.

The consultation paper identifies and describes 16 marine consultation areas around the coast of England and Wales including designated marine nature reserves. Promoters of schemes that may affect a marine consultation area should consult either English Nature or the Countryside Council for Wales as appropriate at the earliest stage, and maintain dialogue at all stages of development. Bodies responsible for giving permission for the development should not only ensure that the consultation has taken place, but should also take the advice of English Nature or the Countryside Commission for Wales into account when reaching a decision. The success of the scheme would be monitored through annual reports, which would be available to the public on request. A similar scheme is already in operation in Scotland.

Table 5.2 Summary of UK legislation affecting water supply and wastewater treatment schemes

Date	Legislation	Purpose
Acts		
1949	Coast Protection Act	Makes provision for *inter alia* navigation safety. Consequently consent is required from the Department of Transport prior to building structures such as sea outfalls
1949	National Parks and Access to the Countryside Act	Designation of national parks, Sites of Special Scientific Interest (SSSI) and nature reserves
1956	Clean Air Act	Concerns air pollution and measures to abate it through adjustments to chimneys and the designation of smoke control areas. Some of the provisions of this Act have been superseded by EPA Part I
1968	Clean Air Act	Makes provision for abating air pollution and covers emissions of smoke, grit and fumes from industrial or trade premises. Some of the provisions of this Act have been superseded by EPA Part I
1968	Sewerage (Scotland) Act	Makes provision for wastewater treatment in Scotland
1974	Control of Pollution Act	The Act is concerned with the collection and disposal of waste with a requirement for all disposal sites that take controlled wastes to be licensed
1980	The Water (Scotland) Act	Makes provision for water supply in Scotland
1981	Wildlife and Countryside Act	The Act is concerned with the protection of wild animals, plants and habitats, conservation of the countryside and protection of public rights of way
1985	Food and Environment Protection Act	Licences are required to construct a long sea outfall
1985	Wildlife and Countryside (Amended) Act	SSSI notification
1989	Water Act	Establishes NRA as a regulatory body with responsibilities for the control of pollution and water resources, and the water companies with responsibility for water supply and sewage treatment
1990	Environmental Protection Act	Part I introduces integrated pollution control (IPC) regulated through HMIP and local authorities for air pollution control. Operators are required to implement BATNEEC. Centrally controlled processes likely to involve emissions to more than one environmental medium need to be implemented using BPEO
		Part II covers waste on land
		Part III covers statutory nuisances, e.g. noise, odour, dust, smoke and steam
		Part VII concerns the replacement of the Nature Conservancy Council with three regional councils for England, Wales and Scotland
1990	Town and Country Planning Act	Makes provision for planning control through the adoption of structure plans and local plans, and gives local planning authorities the power to grant or refuse planning permission
1991	Water Resources Act	Defines functions of the NRA, deals with water resources management, and the control of the pollution of water resources including the issue of abstraction and impoundment licences and discharge consents to controlled waters
1991	Water Industry Act	Defines functions of water and sewerage undertakers, the Director General of Water Services and local authority with regard to water supply. Includes duty on water suppliers to supply wholesome water and issue of consents to discharge trade effluent to the public sewer
1991	Statutory Water Companies Act	Defines the functions of the statutory water companies
1991	Land Drainage Act	Describes the functions of the inland drainage boards and land drainage functions of local authorities
1991	Water Consolidation (Consequential Provisions) Act	Covers amendments and repeals from earlier statutes
1991	Planning and Compensation Act	Parts I and II amend the law on town and country planning in England, Wales and Scotland
1991	Roads and Street Works Act	Concerns procedures for undertaking road works including notices, coordination of works, public information registers, special controls, reinstatement and need for training operatives

continued/..

Table 5.2 (continued)

Date	Legislation	Purpose
Regulations and orders		
1988	Town and Country Planning (Assessment of Environmental Effects) Regulations SI No 1199	Sets out the environmental assessment procedure in line with Council Directive 85/337/EEC for projects requiring planning permission in England and Wales. See also General Development Order 1988 (Article 14(2) SI No. 1813
1988	Environmental Assessment (Scotland) Regulations SI No 1221	As above for Scotland
1989	Planning (Assessment of Environmental Effects) Regulations (Northern Ireland) SR No 20	As above for Northern Ireland
1989	The Air Quality Standards Regulations SI 317	Enforces air quality guide and limits values for sulphur dioxide and suspended particles as given in Council Directive 80/779/EEC, limits on airborne lead as given in Council Directive 82/884/EEC, and nitrogen dioxide as given in Council Directive 85/203/EEC
1989	Sludge (Use in Agriculture) Regulations SI No 1263 — amended SI 1990/880	Sets limits on heavy metals in sludges, application and agricultural practices in line with Council Directive 86/278/EEC
1989	Water Supply (Water Quality) Regulations SI No 1147 — as amended by SI 1989/1384, SI 1991/1837 and SI 1991/2790	Sets water quality standards, monitoring procedures and treatment methods
1989	Surface Waters (Classification) Regulations SI No 1148	Sets water quality criteria by which to classify surface waters in terms of their suitability for abstraction as drinking water giving effect to Council Directive 75/440/EEC
1989	Trade Effluents (Prescribed Processes and Substances) Regulations SI No 1156 — amended SI 1990/1629	Sewerage undertakers have to refer to the Secretary of State the decision as to whether to prohibit or impose conditions on trade effluent discharges containing substances or resulting from processes listed in these Regulations
1989	Surface Waters (Dangerous Substances) (Classification) Regulations SI No 2286	Sets out a classification system for inland waters, coastal waters and territorial waters based on concentrations of dangerous substances listed. The classifications are required for setting water quality objectives under the Water Act 1989 and take account of water quality standards in EC Directives 82/176, 83/513, 84/156, 84/491 and 86/280
1990	Nitrate Sensitive Areas (Designation) Order SI No 1013	Defines nitrate sensitive areas in England within which farmers may enter into an agreement to limit the use of nitrogen fertilizer
1990	The Water Supply (Water Quality) (Scotland) Regulations SI No 119	Corresponds to Regulations SI 1989/1147 above
1990	Surface Waters (Classification) (Scotland) Regulations SI No 121	Corresponds to Regulations SI 1989/1148 above
1990	Surface Water (Dangerous Substances) (Classification) (Scotland) Regulations SI No 126	Corresponds to Regulations SI 1989/2286 above
1991	The Environmental Protection (Prescribed Processes and Substances) Regulations SI No 472	Lists processes prescribed under EPA, establishing whether they come under HMIP or local authority control and prescribed substances released into the air, water or onto land

continued/...

Table 5.2 (continued)

Date	Legislation	Purpose
1991	The Environmental Protection (Applications, Appeals and Registers) Regulations SI No. 507	This sets out the requirements for applications for an authorisation under EPA Part I, the appeal procedures and the keeping of registers
1991	The Bathing Waters (Classification) Regulations SI No 1597	Gives legal backing in England and Wales to Council Directive 76/160/EEC to classify bathing waters
1991	Bathing Waters (Classification) (Scotland) Regulations SI No 1609	Corresponds to Regulations SI 1991/1597 above
1991	Private Water Supplies Regulations SI No. 2790	Governs water quality from private supplies in England and Wales for drinking, washing or cooking or for food production, supplementing Part III of the Water Industry Act 1991 and implementing Council Directive 80/778/EEC
1992	The Surface Waters (Dangerous Substances) (Classification) Regulations SI No 337	Implements Council Directive 86/280/EEC as amended by 90/415/EEC for four substances — ethylene dichloride, tri- and perchloroethylene and tri-chlorobenzene — which will be incorporated into the SWQOs under the Water Resources Act 1991
1992	The Surface Waters (Dangerous Substances) (Classification) (Scotland) Regulations SI No 574	Corresponds to Regulations SI 1992/337 above
1992	The Private Water Supplies (Scotland) Regulations SI No 575	Corresponds to Regulations SI 1991/2790 above
1992	The Controlled Waste Regulations SI No 588	One of the items of concern is that sewage and septic tank sludges disposed of to agricultural land under Regulations SI 1263 are not to be treated as industrial or commercial wastes under EPA Part II

In the UK, sites are designated for reasons other than nature conservation: for example, landscape, archaeological, cultural or historic reasons. Such sites have protection either through specific legislation or through the adoption of conservation policies in local and structure plans. They include national parks, areas of outstanding natural beauty (AONB), heritage coasts, regional parks, archaeological sites including scheduled ancient monuments (SAM), listed buildings, and tree preservation orders.

5.2.3 Legislation on water supply and pollution control

In the last four years there have been two major developments in UK legislation. These concern the overhaul of legislation for the water industry and the implementation of the Environmental Protection Act 1990. The following is a brief summary of these developments.

The Water Act 1989 separated the functions of the old water authorities in England and Wales and led to the privatisation of the water industry. The National Rivers Authority (NRA) was established as a regulatory body with responsibilities for the control of pollution in freshwaters, groundwaters and coastal waters, and the management of water resources. It was also given responsibilities for flood defence, land drainage, fisheries, navigation, nature conservation and recreation in inland waters and associated lands. The water authorities' responsibilities for water supply and wastewater treatment were given over to the newly privatised water service companies in England and Wales. Water undertakers, including both the existing water supply companies and the water service companies, were given responsibilities for nature conservation and recreation on their land. They also monitor drinking water quality, but the responsibility for checking lies with the local authorities and Drinking Water Inspectorate.

In 1991 five Acts were introduced (Table 5.2) consolidating legislation regarding the water industry. From the point of view of this chapter the two most important ones are the Water Resources Act and the Water Industry Act. The former consolidated the powers of the NRA while the latter concerned water supply and wastewater treatment.

The Water Resources Act 1991 defines the NRA and its duties (Part I), deals with water resources management (Part II) and the control of the pollution of water resources (Part III) among other things. The NRA safeguards water resources through the issue of abstraction and impoundment licences and discharge consents to controlled waters. Under the legislation the Secretary of State has powers to:

- classify all controlled waters according to water quality
- set statutory water quality objectives (SWQOs) for those waters, and
- prevent and control pollution including designating water protection areas and nitrate-sensitive areas.

Water quality protection is based on identifying the legitimate use of these waters and setting SWQOs. The water quality objective would be met if the water quality meets the prescribed water quality standards, which include Council Directives on water quality (Table 5.3).

The discharge consent stipulates, among other things, the location of the discharge; the design and construction of any outlets; the nature, composition, origin, temperature, volume and rate of discharge; and times at which discharges may be made. The NRA may request data or an EA from an operator in order to determine the details of a discharge consent. As a result the discharge consent is designed to avoid the deterioration of water quality to a level below the relevant quality standard.

Under the Water Industry Act 1991 water suppliers have a duty to supply wholesome water; local authorities have a duty to keep themselves informed about the wholesomeness and sufficiency of supply; and operators require a consent from the sewerage undertakers to discharge trade waste to the sewerage system.

A series of Council Directives have either set water quality standards for different uses of water, or prescribed the control of emissions of certain substances. In the former case there are for example Directives covering the quality of water for drinking and human consumption (75/440/EEC, 79/869/EEC and 80/778/EEC), bathing waters (76/160/EEC), and fish life in fresh waters and shellfisheries (78/659/EEC and 79/923/EEC). In the latter case the Dangerous Substances Directive (76/464/EEC) is aimed at preventing the discharge of some pollutants and limiting the discharge of others; however, there are many other Directives in this category (Table 5.3). These Directives are being given statutory backing in the UK through the adoption of Statutory Instruments (Table 5.2).

The Water (Scotland) Act 1980 and Sewerage (Scotland) Act 1968 rested responsibilities for water supply and wastewater treatment with the 12 Regional and Islands Councils, while responsibility for pollution control lies with the seven River Purification Boards and the three Islands Councils. In Northern Ireland the protection of water quality, water supply and wastewater treatment is the responsibility of the Department of the Environment for Northern Ireland.

Table 5.3 Summary of European Community Council Directives affecting water supply infrastructure and wastewater treatment schemes

Type of scheme	Date	Council Directive	Purpose
Environment	1979	On the conservation of wild birds (79/409/EEC)	To maintain populations of naturally occurring wild birds and to preserve a sufficient diversity and area of habitat
	1985	On the assessment of the effects of certain public and private projects on the environment (85/337/EEC)	Defines projects requiring environmental assessments (EA), the scope of EAs, procedures for consultation with the public and planning application procedures
	1990	On the freedom of access to information on the environment (90/313/EEC)	Requires information on environmental issues held by public authorities to be available on request to individuals under certain conditions
	1992	On the conservation of natural habitats and of wild fauna and flora (92/43/EEC)	To identify and designate sites of community importance for the conservation of habitats and species of flora and fauna
Water quality	1975	On the quality of surface water for drinking (75/440/EEC)	Sets quality requirements for surface water intended for the abstraction of drinking water
	1975	On the quality of bathing water (76/160/EEC)	Sets requirements for water quality parameters in fresh or sea water where bathing is explicitly authorised by the authorities or is traditionally practised by a large number of bathers
	1976	On pollution caused by the discharge of certain dangerous substances into the aquatic environment (76/464/EEC)	Identifies substances that must be eliminated from discharges to inland surface waters, territorial waters, internal coastal waters, and groundwater (List I) and those that must be reduced in discharges to the same aquatic environments (List II)
	1978	On the quality of fresh waters needed to support fish life (78/659/EEC)	Sets water quality standards for salmonid and cyprinid waters designated as requiring protection or improvement
	1979	On the quality required of shellfish waters (79/923/EEC)	Sets water quality standards for coastal and brackish water in designated shellfisheries
	1979	On the sampling and analysis of surface water for drinking (79/869/EEC)	Sets methodology to be used to measure the parameters given in Council Directive 75/440/EEC
	1979	On the protection of groundwater against pollution caused by certain dangerous substances (80/68/EEC)	Seeks to prevent the pollution of groundwaters by List I substances and to limit the pollution of groundwater by List II substances
	1980	On the quality of water for human consumption (80/778/EEC)	Sets water quality standards for all water intended for human consumption, including that used in the food industry
	Various	73/404/EEC, 78/176/EEC, 82/176/EEC, 82/883/EEC, 83/513/EEC, 84/156/EEC, 84/491/EEC, 86/280/EEC, 89/428/EEC, 90/415/EEC	Directives setting limit values and quality objectives for substances discharged to the aquatic environment: detergents; titanium oxide waste; mercury from chlor-alkali electrolysis industry; titanium oxide industry waste; cadmium; mercury (other industries); hexachlorocylo-hexane; substances in List 1 of 76/464/EEC; titanium dioxide industry waste; an amendment to 86/280/EEC.
Wastewater treatment and sludge disposal	1986	On the protection of the environment when sewage sludge is used in agriculture (86/278/EEC)	Sets limit values on heavy metal concentrations in sewage sludge applied to agricultural land and the soil; methodologies for undertaking sludge soil analysis; procedures for applying sludge; and methods for recording information
	1991	Concerning the protection of waters against pollution caused by nitrates from agricultural sources (91/676/EEC)	To reduce water pollution caused by nitrates from agricultural sources and to prevent further such pollution
	1991	Urban waste water treatment (91/271/EEC)	Sets requirements for the collection, treatment and disposal of urban waste water and the treatment and disposal of waste water from certain industrial sectors

continued/...

Table 5.3 (continued)

Type of scheme	Date	Council Directive	Purpose
Incineration and air quality	1980	On air quality limit values for sulphur dioxide and suspended particulates (80/779/EEC)	Sets limit and guide values for sulphur dioxide and suspended particulates in the atmosphere
	1984	On the combating of air pollution from industrial plants (84/360/EEC)	Requires prior authorisation for the operation of certain plants (including liquid and solid waste incinerators) and that pollution control measures are used following BATNEEC. Defines most important polluting substances
	1985	On air quality standards for nitrogen dioxide (85/203/EEC)	Fixes a limit value and guideline values for nitrogen dioxide in the atmosphere
	1989	On the prevention of air pollution from new municipal waste incineration plants (89/369/EEC)	Excludes plants used specifically for the incineration of sewage sludge. Sets requirements on emissions and operating procedures before authorisation can be granted to new plants
	1989	On the reduction of air pollution from existing municipal waste incineration plants (89/429/EEC)	Excludes plants used specifically for the incineration of sewage sludge. Sets standards and operating procedures which have to be adapted in existing plants by various deadlines
Other Directives	Draft	Landfill	To determine the procedures for granting permits; site selection and operation; reporting requirements; pricing; liability; and closure and aftercare
		Incineration of hazardous waste	Considers the design, operation and emissions standards of incinerators. List of hazardous wastes not completed. Excludes municipal waste incinerators

Table 5.4 International Conventions on conservation

Date	Convention	Purpose
1975	Convention on wetlands of international importance especially on waterfowl habitat – The Ramsar Convention.	Designates wetlands for inclusion in a list of wetlands of international importance and promotes conservation of these wetlands
1975	Convention concerning the protection of the world cultural and natural heritage – World Heritage Convention.	Protection of sites of international cultural and natural heritage
1979	Convention on the conservation of European wildlife and natural habitats – Berne Convention.	Maintains populations of wild flora and fauna, with particular emphasis on endangered and vulnerable species
1983	Convention on the conservation of migratory species of wild animals – Bonn Convention.	Protects migratory species including birds, mammals, fish and invertebrates

The Environmental Protection Act 1990 is set out in eight parts, of which the following four sections are most relevant to this chapter.

- Part I – Integrated Pollution Control (IPC) and Air Pollution Control by Local Authorities
- Part II – Waste on Land
- Part III – Statutory Nuisances and Clean Air
- Part VII – Nature Conservation in Great Britain and Countryside Matters in Wales.

Under Part I of EPA certain processes and substances released to the atmosphere, land or water are prescribed by regulations. Operators have to have a licence to use the prescribed processes or discharge the prescribed substances. These are obtained from HMIP for centrally controlled processes and emissions to air, water and land, or from local authorities for locally controlled processes and emissions to the atmosphere only. The Environmental Protection (Prescribed Processes and Substances) Regulations 1991 establishes whether control comes under HMIP or the local authority. One of the objectives of the licence is to ensure that the 'best available techniques not entailing excessive costs' (BATNEEC) are used to prevent or reduce emissions and to render substances harmless. For processes that are centrally controlled (under IPC) and are likely to include emissions to more than one medium (i.e. land, water and air) the 'best practicable environmental option' (BPEO) would be followed.

At present there is a potential overlap between HMIP and NRA regarding emissions to water. The organisations approach pollution control from a different perspective, with HMIP looking at the control of the source of emissions through BATNEEC and the NRA looking at water use and working backwards to determine acceptable levels of emissions. To overcome potential duplication the NRA and HMIP have produced a memorandum of understanding.

HMIP is producing guidance notes for their inspectors and have published one on sewage sludge incineration (Process Guidance Note IPR 5/11). In the longer term this may be governed by a council directive on hazardous waste incineration which is still in the draft stage (Table 5.3). The list of hazardous waste has not yet been finalised so it is not clear whether this would include sewage sludges. Municipal waste incinerators are already covered by council directives.

Part II of the EPA 1990 concerns the regulation of waste disposal on land and interacts with the Control of Pollution Act (1974). Three new authorities have been established:

- waste regulation authorities (WRA) with responsibilities for granting waste disposal licences and subsequent control

- waste disposal authorities (WDA) responsible for the provision of disposal facilities

- waste collection authorities (WCA) responsible for arranging the collection of waste.

Local authority waste disposal companies (LAWDC) are being set up to undertake waste collection and disposal. The Act also sets out measures for improving the operation and control of waste disposal sites and imposing a duty of care, all of which has implications for the disposal of sludges, incinerator ashes and industrial wastes to landfill.

A draft Council Directive on landfill may affect implementation of Part II of the EPA 1990, although it is difficult to assess the extent of the impact of the Directive on UK legislation until the final text has been adopted. The present draft distinguishes between hazardous waste, municipal waste and inert waste. It seems that liquid wastes and sludges could be disposed of at municipal (non-hazardous) disposal sites provided compatibility criteria set out in Annex III are met. However, whether ash from incinerators would be considered as a hazardous waste would presumably depend on its chemical composition, particularly with regard to heavy metal concentrations.

Part III of the EPA 1990 deals with the definition of statutory nuisances; the inspection of premises and summary proceedings for statutory nuisances in England and Wales; and an amendment to the Public Health (Scotland) Act 1897 on the definition of nuisances. Statutory nuisances include dust, steam, smell and other effluvia arising from industrial, trade or business properties and noise emissions which are prejudicial to health or a nuisance. Complaints about water and wastewater treatment works (including sludge dryers, which are not prescribed under IPC, unlike sludge incinerators) would be dealt with under this part of EPA and, where applicable, the Clean Air Acts (1956 and 1968) and the Control of Pollution Act 1974.

Part VII of the EPA 1990 concerns the division of the Nature Conservancy Council into new councils for England, Scotland and Wales, now referred to as English Nature, the Nature Conservancy Council for Scotland and the Countryside Council for Wales.

Other legislation pertinent to pollution control concerns urban wastewater treatment, the application of sewage sludges on agricultural land, nitrate pollution from agricultural land, and air pollution from incinerators.

The Council Directive on urban wastewater treatment (91/271/EEC) sets dates by which settlements of certain sizes need to be served by a collecting system, levels of treatment required for wastewater treatment works, quality standards for treated effluent, sampling procedures for the treatment of urban wastewater, and the cessation of dumping sewage sludges at sea by 1998. The criteria for determining levels of wastewater treatment vary according to the size of the urban agglomeration and the nature of the receiving waters. Secondary or equivalent treatment is required by 31 December 2000 for wastewater discharges from agglomerations greater than 15,000 pe (population equivalents), except where:

- the receiving waters are designated 'sensitive areas', in which case a higher level of treatment is required for all agglomerations greater than 10,000 pe

- the receiving waters are coastal waters defined as 'less sensitive areas', in which case urban wastewater from agglomerations with 10,000 – 150,000 pe may be subjected to less stringent treatment, provided that discharge receives at least primary treatment, and that comprehensive studies indicate that the discharges will not affect the environment adversely.

EC member states have to identify 'sensitive' and 'less sensitive areas' by 31 December 1993 following the criteria used in Annex II of the directive. Sensitive areas would fall into one of three groups:

- waters which are, or may in the near future become, eutrophic
- surface freshwaters intended for abstraction for drinking waters
- areas where more than secondary treatment is required to satisfy council directives.

A 'less sensitive area' is one where the marine water body is not adversely affected by the discharge of wastewater. In the UK, neither the 'sensitive' nor 'less sensitive areas' have been designated at present. However, the Department of the Environment has issued criteria and procedures for identifying sensitive and less sensitive areas for consultation.

As a result of this legislation, treatment works are being constructed to provide at least primary treatment where previously untreated wastewater was discharged to sea. Other treatment works located in sensitive areas are being upgraded to provide a higher level of treatment. One of the consequences of this action is to increase greatly the quantity of sewage sludges produced. The main disposal route, the application of sewage sludges to agricultural land, is controlled by Council Directive 86/278/EEC, which has been given statutory backing in the UK through the Sludge (Use in Agriculture) Regulations 1989. Runoff from agricultural land is thought to be one of the main sources of nitrates in surface and groundwaters as a consequence of the application of fertilisers. This has led to the designation of nitrate sensitive areas (NSA) under the Nitrate Sensitive Areas (Designation) Order SI No 1013 (1990) and the Council Directive on nitrates (91/676/EEC). In the UK ten NSAs have been designated, within which farmers undertake agricultural practices to reduce nitrate leaching.

Finally there are a number of legislative instruments dealing with air pollution in addition to IPC. Those that may have a bearing on the drying and incineration of sewage sludges are described briefly in Tables 5.1 and 5.2.

5.2.4 Legislation concerning construction in the water industry

There are laws governing certain aspects of construction in the water industry. For example, a licence is required from the NRA to impound a river under the Water Resources Act 1991; the construction of sea outfalls is affected by several pieces of legislation (ICE, 1989); and works involving digging up streets are governed by the Roads and Street Works Act 1991. These examples of legislation are considered briefly in Table 5.2.

5.3 PRINCIPAL EA AND CONSTRUCTION ACTIVITIES

5.3.1 Environmental Assessment methodology

The aim of an EA is to determine the likely effects of a development on the surrounding environment during the construction, operation and decommissioning stages of the scheme. It is an iterative process, which requires interaction between environmental assessment and engineering design, from the initial feasibility stage, through to detailed design. In this way environmental constraints are considered at an early stage of the project. The EA is often seen to finish with the production of the environmental statement (ES), but following a best-practice approach the assessment should continue throughout the life of the project.

The main steps involved in a best practice approach to EA are:

- scoping study
- collection of existing data
- field surveys
- consultation
- assessment of scheme and alternative options
- prediction of impacts
- design of mitigation measures
- presentation of results in an ES
- monitoring
- management agreements
- review.

A scoping study is a short, initial review aimed at a preliminary assessment of the nature of the site, and the most likely positive and negative effects of the scheme on the environment. This may be carried out by a consideration of the proposed scheme, field visit, literature search and informal discussions with consultees. The last provide not only useful data, but often insight into the nature of potential impacts based on personal knowledge of the area and a first airing of possible objections from the consultees.

In order to assess the impacts of a scheme, it is necessary first to characterise the existing conditions. This proceeds in two directions: the collection and study of existing data, and the commissioning of 'baseline' surveys. Information about a site may be collected from many sources. Published maps provide national coverage of a large number of parameters useful for background information such as:

- topography, numbers and location of properties, lines of communications (Ordnance Survey maps at various scales)

- soil classification (Soil Survey of England and Wales)

- agricultural land use and farm type (MAFF Publications)

- geology (British Geological Survey)

- bathymetry, seabed materials, current directions and speeds (Admiralty charts).

Reasonably up-to-date aerial photographs are very valuable in locating habitat types, land use, archaeological sites, and geomorphological features. A literature search or consultation with research bodies can prove useful in obtaining site-specific information. Finally, existing data are often available from many types of organisation for a reasonable charge.

These data need to be scanned to determine whether there are any vital gaps in the data sets. From this information it will be possible to determine what types of field survey are required to supply outstanding data. Table 5.5 lists the type of information required for baseline surveys. These surveys need to be carried out following a systematic, replicable methodology so that differences in environmental conditions found by later surveys can be attributed to the development or other factors rather than methodological errors.

Consultation is a legal requirement to ensure that the public are fully informed of the effects of a development and have an opportunity to discuss the scheme, its impacts and mitigation measures before a planning decision is made. A list of potential consultees is given in Table 5.6. In following a best-practice approach it is preferable to consult the statutory and main non-statutory consultees as early as possible and to maintain contact throughout the assessment period. This ensures that objections to a proposal are brought to the attention of the developers early in the design stage when modifications, or alternatives, can be more readily developed. Consultees often hold data on the site and may offer advice on baseline surveys, the significance of impacts, mitigation and monitoring. Early consultation may also avoid confrontation, delay and possible further objections after the proposal is submitted for planning approval. The following is a list of the most likely statutory consultees for water supply infrastructure and sewage treatment projects given in Regulation 8 of the Town and Country Planning (Assessment of Environmental Effects) Regulations 1988:

- local and county authorities

- the Countryside Commission

- English Nature, Countryside Council for Wales, or Nature Conservancy Council for Scotland

- National Rivers Authority

- HMIP

- British Coal Corporation (for a development which affects an area of coal working notified by the British Coal Corporation to the local planning authority)

- Historic Buildings and Monuments Commission or the Secretary of State for Wales (for developments in England and Wales respectively which would affect a listed building or scheduled ancient monument)

- Minister of Agriculture, Fisheries and Food, Secretary of State for Wales or Scottish Office (Agriculture and Fisheries) for projects affecting grade 1, 2, or 3a agricultural land.

The assessment of the proposal also requires that the EA team has a thorough understanding of the proposal. This includes the physical characteristics of the project (such as size and layout), the main construction and operational activities, the type and quantity of raw materials used and the possible emissions. This information is needed to predict impacts and has to be described in the ES (Section 2.2). The main activities for water supply infrastructure and wastewater treatment works which may result in potentially significant impacts are shown in Table 5.7 and described further in Section 5.3.2. Ideally the, proposal should adopt BATNEEC and, where there is a choice, BPEO. It is necessary to examine alternative options (for example, location, site layout, processes or waste disposal options), and to discuss reasons for rejecting any options.

The potential effects of a scheme on the environment need to be determined. These include direct and indirect; secondary; cumulative; short-, medium- and long-term; permanent and temporary; and positive and negative effects. The magnitude of many of the effects can only be assessed qualitatively, although quantitative assessments can sometimes be made. Having determined the type and likely significance of an effect, measures can be designed to mitigate them. A discussion on the identification, evaluation and mitigation of the environmental effects of water supply infrastructure and wastewater treatment works is given in Section 5.4.

Table 5.5 Baseline data requirements

Environmental characteristics	Examples of appropriate data
Flora and fauna	Habitat type; species composition and abundance; rare species; designated sites; breeding grounds
Commercial fisheries	Type, location, economic value
Air quality	Existing levels and sources of pollution
Water quality	Physical, chemical and microbial constituents
Sediment quality	Particle size; cohesion; lithology; characteristics of sludges
Soil/ground conditions	Soil type; geotechnical characteristics; contaminated land
Geology	Lithology; seismology; protected sites
Climate	Wind speed and direction; local atmospheric conditions; rainfall
Hydrology	Discharge; flood and low flows; channel morphology; abstractions; returns
Groundwater	Water quality; recharge flow; abstractions; designated nitrate sensitive areas
Tidal regime	Tides; currents; storm surges; waves; wind
Oceanography	Bathymetry; seabed materials; sediment transport
Population	Density; numbers; structure; trends
Local community	Attitude to development
Housing	Settlement patterns; quality; location
Employment	Major industry types/employers; unemployment rate; trends
Health and safety	Health of population (e.g. incidence of respiratory diseases)
Emergency services	Location and resources
Land use	Type; economic value; planning policies
Agriculture	Grade of agricultural land; farm structure and viability; severance; trends
Landscape	Designated areas; quality; characteristics
Leisure and amenity	Activities; facilities
Tourism	Numbers; seasonality; facilities; attractions
Heritage	Designated sites
Infrastructure	Transport network and capacity; utilities
Traffic	Mode of transport; amount of traffic; percentage HGV on roads; noise; vibration
Navigation	Minimum flow required in rivers
Planning framework	Policies; planning applications
Noise	Existing levels; diurnal variations; tone; sources
Odour	Degree of offensiveness; sources
Dust	Deposition rates; incidents of spoiling; areas affected

Table 5.6 Possible consultees in the EA process in the UK

Environmental parameter	Primary interested agencies
Flora and fauna	English Nature/Nature Conservancy Council for Scotland/Countryside Council for Wales; Royal Society for the Protection of Birds; British Trust for Ornithology; Wildfowl and Wetland Trust; local wildlife and naturalists trusts; National Trust; Royal Society for Nature Conservation; Marine Conservation Society; Forestry Commission; Botanical Society of the British Isles; Scottish Wildlife Trust; National Parks; National Rivers Authority/River Purification Boards
Fisheries	Ministry of Agriculture, Fisheries and Food/Department of Agriculture and Fisheries for Scotland: National Rivers Authority/River Purification Boards; riparian owners; sea fisheries districts; local angling associations; Sea Fish Industry Authority; Fisheries Statistical Unit (Scotland); The Scottish Salmon Fisheries Inspector
Air quality	Local planning authority, HMIP
Water quality	National Rivers Authority/River Purification Boards; HMIP
Sediment quality	Ministry of Agriculture, Fisheries and Food; National Rivers Authority; local waste regulation authority
Soil	Soil Survey of England and Wales; Ministry of Agriculture, Fisheries and Food; Department of Agriculture and Fisheries for Scotland
Geology	British Geological Survey; English Nature/Nature Conservancy Council for Scotland/Countryside Council for Wales
Climate	Meteorological Office
Hydrology/groundwater	National Rivers Authority/River Purification Boards; Inland Drainage Boards; British Waterways; Institute of Hydrology; Institute of Geological Sciences; nature conservancy groups
Tidal regime/coastal configuration/shore	National Rivers Authority; Ministry of Agriculture, Fisheries and Food; Meteorological Office; Marine Information Advisory Service; local port authorities; Department of Transport; nature conservancy groups
Population/local community/housing/local economy/employment	Department of Employment; local chamber of commerce; county council; local planning authority; parish council
Health and safety	Health and Safety Executive; local planning authority (environmental health officer)
Emergency services	Emergency services
Land use/agriculture	Ministry of Agriculture, Fisheries and Food/Department of Agriculture and Fisheries for Scotland; Local Planning Authority; National Farmers Union; Country Landowners' Association; Crown Estate Commissioners; Council for the Protection of Rural England; National Parks Authority; British Coal Corporation; waste disposal authority; landowners
Landscape	Countryside Commission; local planning authority; National Rivers Authority; Council for the Protection of Rural England; National Parks Authority
Leisure and amenity	Local planning authority; tourist board; Sports Council; Ramblers Association; local sports associations; English Heritage; National Parks; National Rivers Authority; water companies
Tourism	Local planning authority; tourist board; Sports Council; county council
Heritage	Historic Buildings and Monuments Commission (English Heritage); Royal Fine Arts Commission for Scotland; Secretary of State for Scotland; county council; county or local museums; local planning authority; National Trust; Landmark Trust; Building Research Advisory Service; National Rivers Authority
Infrastructure	County council; local planning authority; utility companies
Traffic	Department of Transport; county council; local planning authority; British Railways Board; Traffic and Road Research Laboratory
Navigation	Department of Transport; Port Authority; Harbour Commissioners; National Rivers Authority/River Purification Boards; British Waterways

Consideration may also be given to environmental enhancement, that is, additional measures to improve the environment. For example, measures could be taken to improve nature conservation at the site, provide recreational facilities such as footpaths, or replace existing overhead cables with underground ones to improve the aesthetic appearance of a site.

The results of the EA need to be written up in an ES, the content of which is described in the Directive. However, once planning permission has been obtained, the assessment should ideally continue with the setting up of monitoring systems to check on the impact of the scheme during construction and operation. For example, the performance of outfalls or incinerators may be checked by sampling water and air quality and surveying the ecology of the sites. Monitoring systems would give an early warning of pollution, which would allow remedial action to be organised in order to minimise the impacts. They would also provide a check for determining the accuracy of the predictions of impacts set out in the ES. In some cases, management programmes may have been agreed as part of a mitigation scheme, for example the maintenance or aftercare for new plantings. At present these procedures are not routinely carried out.

5.3.2 Summary of main engineering activities: water supply projects

General

Water supply is necessary for agricultural purposes, domestic consumption, power generation, and industrial uses. In 1990 an estimated 35,249 megalitres/day were abstracted in England and Wales with 80% from surface and 20% from groundwater sources (based on returns made by licence holders and estimates by the NRA (DoE, 1992)). Of this total, 52.0% was piped mains water, abstracted by the water service and water supply companies, 46.6% was abstracted directly by industry and 1.4% was abstracted by agricultural users. Of the piped mains water some 70−80% was for domestic consumption and the remainder was used by industrial consumers in the UK.

In order to supply these quantities of piped mains water, it is necessary for the water service companies to abstract water, and sometimes to store it, and transfer it from the source area to the demand area, prior to treatment and distribution to domestic and industrial premises. This may require the construction of a range of structures described in Table 5.1. Water supply schemes can have very widespread and severe environmental impacts due to the size of structures needed, their location in the countryside, where nature conservation is an important consideration, and the size of the area affected which may, in a small number of cases, extend to several river basins.

The main industrial users who abstract water directly are the electricity-generating companies. They accounted for 76.9% of the water abstracted for industrial use in 1990 (12,612 megalitres/day (DoE 1992)), much of which was used for cooling and was returned to the rivers locally. Industries too may have to construct river intake structures or boreholes, pumping stations, water storage reservoirs or tanks, pipelines and treatment works in order to meet their requirements.

In farming, the infrastructure required for water supply may be of relatively small cost. This may involve mounting a pump on a tractor to abstract water from a river, pumping groundwater, or diverting water from a river by building off-take and water conveyance structures. In 1990, 378 megalitres/day (or almost 75% of the water abstracted for agricultural use) were used in spray irrigation (DoE, 1992). This water is generally used during very dry periods and much is lost to the fluvial system owing to evaporation and uptake by crops. River diversions to supply fish farms can also cause low flows or even result in the drying up of the channel over a reach. While the impact of any one farmer's activities in a catchment may be localised and small scale, the cumulative impact of the farming community can have severe and widespread effects on groundwater levels, soil drainage, river flows and water quality.

River impoundments

River impoundments have potentially large impacts on the environment owing to the inundation of extensive areas, and the effects on hydrology and ecology downstream. Plans to develop water supply by building dams start with studies to choose suitable sites, usually based on geotechnical, engineering and environmental grounds, followed by detailed site investigation. It is very important at this early stage to identify sites of conservation value so that they can be avoided altogether, as attempts to recreate habitats are very expensive and success cannot always be guaranteed. Buildings of conservation value and some archaeological sites may be relocated, but they no longer lie in their 'setting'. Some other elements of the archaeological and historical landscape value, such as battlefields and parklands, cannot be relocated.

The main construction activities, listed in Table 5.7, start with the clearance and repositioning of structures such as pylons or buildings and the removal of excess vegetation. This removal, particularly if it has a commercial value such as crops or woodland, allows the realisation of assets, and reduces the amount of organic matter that will be inundated, together with the water quality problems that may arise later due to decomposition. It will be necessary to provide various facilities such as a road access to the dam site, power supply, clean water supply and sewage disposal, temporary and permanent offices, and secure storage areas. Preparing the foundations for the dam often requires activities that are noisy, dusty, cause vibration or are potentially dangerous, such as blasting rock faces, large-scale earthworks and piling. The dam itself may be built out of a variety of materials including rock, concrete, or earth which, if not locally available, require the procurement, transport and on-site storage of very large quantities of raw materials. Control measures are needed to prevent runoff from stockpiles and accidental spillages from polluting nearby streams with high sediment loads and products such as petrol and oils.

Table 5.7a Activities that may result in potentially significant impacts (construction phase)

Activity	Examples	Reservoir	River intakes	Treatment works	Sludge treatment including incineration	Outfalls
Construction phase						
Site works	Clear excess vegetation and buildings; relocate populations; earthworks; store materials	■	□	□	□	□
Dredging	Dredge outfall trench		□			□
Ancillary works	Temporary access; power supplies; water supplies	■	□	□	□	□
Raw materials demand	Aggregate; rock armour to cover outfalls; borrow pits	■	□	□	□	□
Transport of raw materials	Aggregates; concrete; pipes; steel structures	■	□	■	□	■
Transport of employees	Daily journeys to site	□	□	□	□	□
Immigration	Contractors; temporary workforce	□	□	□	□	□
Employment	Local people	□	□	□	□	□
Local expenditure	Materials and services	□	□	□	□	□
Lighting	Construction site	□	□	□	□	□
Vibration	Piling; traffic; earth moving; tunnelling	□	□	□	□	□
Noise	On-site plant; traffic; blasting	■	□	■	■	□
Odours	Sewage and sludges during commissioning			□	□	□
Dust and particulates	Earth moving; stockpiles; blasting	■	□	□	□	□
Gaseous emissions	Exhaust from vehicles	□	□	□	□	□
Aqueous discharges	Pollution incidents; storm runoff	■	□	■	■	■
Dredged material disposal	Trenching		□			□
Solid waste disposal	Spoil; construction material	□	□	□	□	□
Accidents/hazards	Slope failure; spillages	□	□	□	□	□
Navigation	Shipping lanes offshore; river traffic	□	□			□

□ - Occasionally causes significant impact   ■ - Commonly causes significant impact

Table 5.7b Activities that may result in potentially significant impacts (operational and post-operational phases)

Activity	Examples	Reservoir	River intakes	Treatment works	Sludge treatment including incineration	Outfalls
Operational phase						
Structures	Dam; buildings; outfall	■		□	□	□
Maintenance dredging	Protect outfall		□			□
Raw materials demand	Chemicals for water and sludge treatment			□	□	
Transport of raw materials	Untreated sludges			■	□	□
Transport of products	Treated sludges			■	□	
Transport of employees	Daily labour movements					
Immigration	Permanent workforce					
Employment	Long-term job creation					
Local expenditure	Materials and services			□	□	
Lighting	Site works			□	□	
Vibration	Traffic			□	□	
Noise	Traffic noise; centrifuges; power supply	□	□	□	□	
Odours	Sludges and wastewater; transport and treatment			■	■	□
Dust and particulates	Fall out from stack				□	
Gaseous emissions	Emissions from stack				■	
Aqueous discharges	Reservoir releases; disposal of liquid effluent from wastewater treatment works; runoff from sites	■		□	□	■
River flow regime	Attenuation of floods; alleviation of low flows; change in discharge	■	■	□		□
Solid waste disposal	Treated sludges; ash			□	■	
Accidents/hazards	Chemical spillage; fail emissions standards; blockage of outfall			□	□	□
Post operation						
Cessation of activity	Traffic; employment; draw down	□		□	□	
Deterioration of structures	Collapse of structures	□	□	□	□	□
Removal of structures	Physical removal	□	□	□	□	□
Contamination	Sludge storage areas; outfall diffuser head area; chemical handling			□	□	□
Disposal of residues	Contaminated soils			□	□	

□ - Occasionally causes significant impact   ■ - Commonly causes significant impact

The size of workforce required to construct a dam varies according to the size of the project and may range from a few dozen to hundreds. The majority of the labour force is usually drawn from an area within daily commuting distance of the site where available, for example 50 miles, although the resident engineers are often drawn from further afield. As a result there may not necessarily be any need to provide on-site accommodation for large numbers of people; however, this does depend greatly on the size of the workforce and remoteness of the site.

On completion of the dam, the reservoir may take several months to fill depending on the characteristics of the upstream catchment and the weather conditions. It would be necessary to test the equipment, structures and associated buildings. Those involved with the planning, granting of consents and construction for a dam must satisfy themselves that the design is safe as the failure of a dam would have very serious impacts on the downstream channel and floodplain. Once operational there are few labour requirements, although there are legal requirements for periodic checks of the dam, and there is a need for maintenance. However, it is not necessary for permanent staff to be located on site, especially where control of releases is highly automated. Employment opportunities may arise from other developments at the site such as recreation and amenity uses.

There have been few cases of decommissioning in the UK as most reservoirs have not yet reached the end of their economic life. Decommissioning a dam would consist of releasing the water, or pumping it away, creating a permanent outlet and removing unstable structures. The slopes and valley would probably be left to dry out and revegetate naturally.

Pipelines

Water from the reservoir may be used to supplement low flows downstream or in neighbouring catchments in order to sustain abstractions from the river. Alternatively, water may be transferred by pipeline to a water treatment works. The selection of the pipeline route needs to be based on environmental as well as engineering constraints in order to avoid damaging areas of conservation value directly by excavating them, such as archaeological and ecological sites, and indirectly by, for example, affecting field drainage. Excavation is usually by trenching, although the pipeline may be jacked under some structures, and the topsoil and subsoil are stockpiled alongside. This may give rise to dust in the atmosphere and contribute to dirt on roads. The soils and vegetation need to be reinstated carefully with particular attention paid to river crossings, wetland, agricultural land, woodland, hedgerows and other ecological sites. Further discussion on EA procedures for pipelines is presented in Chapter 6.

Intakes

The construction of river intake structures may involve earthworks to construct the intake and measures to divert flow temporarily from the works. Problems during construction may include flooding and bank erosion. Following construction, abstraction may cause problems in downstream water resources and water quality as described in Section 5.4.2.

Water treatment works

The water treatment works may vary in size considerably. The design of the works and hence construction and operation activities will depend on the type of treatment carried out. The most basic level of treatment usually involves disinfection, for example by chlorine, and possibly slow sand filtration. Other forms of treatment being used include ozone and new techniques such as granular activated carbon filters. Many of the construction and operation activities are similar to those described in Section 5.3.3 for wastewater treatment works.

Water supplies

Water supply can be significantly improved through leakage control, an important component being the maintenance and renovation of the water supply system. In environmental terms improving the efficiency of the existing system is preferable to the construction of new reservoirs.

5.3.3 Summary of main engineering activities: wastewater treatment works

General

Wastewater may be discharged to surface waters from non-point sources, such as runoff from land during storms, and from point sources such as drainage and effluent outfalls. Direct discharges from non-point sources to receiving waters are uncontrolled and can have considerable impacts on water quality and ecology. In particular, runoff from agricultural land can be detrimental, but pollution can be controlled to some extent through agricultural practices.

Wastewater treatment systems for domestic properties not connected to the sewerage system include cesspits, septic tanks and small package treatment plants. The latter two can result in surface water discharges and therefore require a consent from the NRA. All three need periodic emptying, with the sludges being tankered to a treatment works. These schemes serve less than 10% of the population, have low infrastructural requirements and cause minor localised environmental impacts. For these reasons they are not considered further.

Industrial premises need consents to discharge directly to the receiving waters, or into the sewerage system. Appropriate treatment has to be provided in order to maintain the quality of effluent specified in their discharge consents, which may involve the installation of a wide range of treatment processes on site.

At industrial sites where wastewater effluent is treated, the most common methods are:

- temperature control
- screening
- neutralising the pH
- filtration (including biological methods)
- settlement (including the use of flocculants)
- aeration
- oil and grease removal
- absorption.

The treated effluent may then be transferred to a wastewater treatment works or discharged directly to surface waters. Many industries that discharge untreated waste are coming under pressure to provide treatment. This has economic implications for the industry as well as posing logistical problems such as the lack of space at urban sites.

Treatment works

Effluent from the sewerage systems normally passes through a wastewater treatment works prior to discharge into the receiving waters. The design of the works will depend to a large extent on the type of treatment carried out: this is categorised as preliminary, primary, secondary and tertiary treatment. An estimated 80% of sewage collected in England and Wales receives primary and secondary treatment (DoE, 1992) including most inland works which discharge to freshwaters. However, many discharges at coastal sites receive lower levels of treatment or none at all. Sewage sludges arising from treatment may be disposed of in a number of ways. Currently some 50% is spread on agricultural land, 16% goes to landfill, 5% is incinerated and 29% is disposed of by other means, mainly to sea.

Construction programmes for wastewater treatment works consist of upgrading and extending existing works as well as building new works at green field sites. Many of these schemes are small scale, but there are some large ongoing projects. For example, the Mersey Estuary Pollution Alleviation Scheme (MEPAS), due for completion in 1996, involves the installation of a 14 mile-long interceptor sewer, the closure of 26 crude outfalls to the Mersey Estuary, and the completion of a new wastewater treatment works at Sandon Dock.

At a wastewater treatment works the following treatment may be provided:

- screening and grit removal (preliminary treatment)

- primary settlement to remove suspended solids, which form a sludge, from the liquid effluent

- biological or chemical processes to remove organic matter (secondary treatment)

- filtering, screening or straining processes to remove fine particulate matter (tertiary treatment).

Treated effluent from the plant would be discharged through an outfall to inland waters or the sea.

The sludges produced by primary and secondary treatments can themselves be treated further in order to reduce the volume, odours and the content of pathogens. The main types of operations are:

- sludge thickening operations involving the use of flocculants
- mechanical dewatering using presses or centrifuges
- aerobic or anaerobic digestion.

There is also considerable research on novel treatment methods that produce usable by-products: for example, soil conditioners through composting and fuel products such as oil-from-sludge (OFS).

Sludge disposal is a major issue as wastewater treatment works are being upgraded and disposal of sludges to sea is phased out as a response to various Directives, particularly the Urban Wastewater Directive and Bathing Water Directive. Some areas, notably Scotland, Thames region, Liverpool, Teeside and Tyneside depend heavily on sea disposal and need to develop alternative disposal strategies. Disposal to land includes application to agricultural land (which will continue to remain a very important disposal option), use in forestry and reclamation works or landfill. The sludges may also be incinerated in dedicated incinerators or with municipal solid waste. The ashes, which may contain dangerous substances such as heavy metals, would then be disposed of in landfill sites. Frost *et al.* (1990) describe sludge treatment and disposal options together with their environmental impacts while Powlesland and Frost (1990) examine BPEO for sludge disposal. In some areas, notably Yorkshire and Thames, the water service companies are developing incineration as a waste disposal option owing to large quantities of sludges arising from urban centres, insufficient agricultural land or landfill sites, the high pollutant levels in sludges from industrialised catchments and high transport costs for moving sludges. In other areas options for disposal to land are being developed.

Wastewater treatment works are normally constructed close to a watercourse (often downstream from the catchment to reduce pumping costs) or, in coastal areas, by the sea, in both cases to facilitate the disposal of effluent. The selection of a new site may be hindered by the need to have sufficient area, while avoiding sites of conservation or recreation value. Sometimes the public are vociferous about the siting of new works close to urban areas, particularly if the work receives sewage from another district.

Construction of a site can be difficult: for example, where space is limited, or where deep underground works are required to pump sewage to ground level. The installation of the sewerage system may involve extensive or deep excavations through populated areas, or difficult terrain such as low lying or hilly areas. Such excavations may result in disruption to traffic, people and farming as well as the temporary storage of spoil and disposal of excess to landfill or recycling for construction. Unless proper precautions are taken for containment of fuels and other hazardous materials, accidental spillages during construction can cause pollution. The construction labour force would vary in size depending on the nature of the project. On-site facilities required would include a secure works compound, a site office and water and power supplies.

During operation, larger wastewater treatment works would be permanently staffed, especially if complex treatment processes are involved. At smaller works, staff would be required to make periodic visits to check on the operation of the plant and quality of effluent, and for routine maintenance. Failure of the treatment works during operation, for example by excessive inputs of certain industrial wastes, can cause severe water quality problems in the receiving waters.

At many small sewage treatment works the sludge may be stored on site temporarily before being emptied into tankers and carried to a regional sludge collecting point for further treatment. Site traffic may also arise from transporting chemicals used in treatment processes, and the disposal of screenings from the preliminary treatment process.

The most common method of decommissioning a treatment works would probably begin with rerouteing wastewater to another treatment works with sufficient capacity. The site should then be cleaned up as soon as possible with the dismantling of structures, and recycling or disposal of all materials and contaminated soils prior to redevelopment.

Sludge

The construction of the buildings required to house sludge and incineration plants does not pose any unusual problems. Consideration needs to be given to odour control and ventilation. The incinerators themselves need to be installed carefully by skilled staff to ensure that they work efficiently. These buildings may be relatively large, being two to three storeys high plus a chimney stack. During operation the plant requires a permanent, skilled workforce to ensure effective operation and maintenance. Decommissioning an incineration plant would require the rerouteing of wastes to another plant, dismantling the structures and cleaning up the sites, including the disposal of contaminated soils.

Outfalls

The construction of an outfall to inland waters would require excavation and the installation of a pipeline. The discharge outfall would need to be located at a point where good mixing of waters occurs, in order to ensure sufficient dilution to mitigate the impact on water quality and ecology. It also needs to be protected from river bank erosion by revetting the surrounding bank. The route would need to avoid sites of conservation interest, as well as other mains and services. The location of the outfall site also has implications for water resources in the locality so that water is recycled rather than being lost to the local system.

The construction of a sea outfall is more complex, owing to the behaviour of inshore waters and nature of the substrate, and is usually restricted to the summer months because of the weather conditions. It would be preferable to avoid an alignment through hard strata, which would require blasting or tunnelling. In sandy substrate the trench can be dredged. The pipe may be laid by one of several methods: for example, by floating and sinking sections, a bottom pull method, pipelaying from a barge or by divers. The method chosen to some extent depends on the type of material used for the outfall (Smy, 1989). Once installed, the pipeline has to be protected from scour and dragging anchors by covering it with rock armour. Any excess debris should be cleared from the seabed to protect fishing equipment.

Several guidelines have been published on outfall site selection, outfall design and environmental protection: for example, Neville-Jones and Dorling (1986), Dorling (1988) and IPHE (1986). Outfalls operate most efficiently when the flow is sufficient to avoid settlement in the pipeline or the ingress of saline water through the diffuser head. They need to be checked and maintained periodically to ensure that they are operating properly.

Outfalls have an expected life of 40–50 years. Decommissioning would involve the cessation of flows, and the dismantling and disposal of the portion of the outfall crossing the foreshore and intertidal areas. It may be impractical to retrieve sections of the outfall buried in deeper waters, although the diffuser heads could be dismantled to reduce any risk to shipping.

5.4 IMPACT IDENTIFICATION, EVALUATION AND MITIGATION

This section considers the potential impacts on a range of parameters, as defined in Table 5.8 below. Owing to the different nature of the impacts involved, water supply infrastructure schemes and wastewater treatment works are considered separately.

Table 5.8 Summary of impacts considered

Main category	Impact	Section No.	Table No.
Physical and chemical effects	Air quality	5.4.1	5.9
	Water quality	5.4.2	5.10
	Water resources	5.4.3	5.10
	Land	5.4.4	5.11
	Physical processes	5.4.5	5.12
Natural habitats		5.4.6	5.13
Human characteristics	Community structure	5.4.7	
	Health and safety	5.4.8	5.14
	Socio-economics	5.4.9	5.15
	Leisure and amenity	5.4.10	5.16
	Land use and landscape	5.4.11	5.17
	Noise	5.4.12	
	Traffic and navigation	5.4.13	
	Cultural heritage	5.4.14	5.17

5.4.1 Air quality

Water supply infrastructure schemes

The storage of large quantities of aggregate on site, haulage along unmade roads and other construction activities at reservoir sites may result in increased levels of dust in the atmosphere on windy days, which may cause a nuisance to nearby residents. The entrainment of dust from spoil heaps could be modelled, with the results checked against on-site dust monitoring. Mitigation measures to reduce the problem include periodically wetting the surface of stockpiles and unsurfaced roads, and putting wind breaks around, or covering the stockpiles.

Changes in microclimate have been observed around large reservoirs, particularly in semi-arid and arid countries where potential evapotranspiration rates are high and air humidity is normally low. In the UK there may still be elevated levels of water vapour in the atmosphere that may, for example, increase the frequency of fog and mists. The potential impact could be estimated by calculating the water balance with and without the reservoir.

Wastewater treatment works

Odour and stack emissions are the two most important impacts of treatment works (including sludge dryers and incinerators). Aerosols may also cause concern.

Odour may be a problem during certain processes (for example cleaning screens or emptying tanks), as a result of changes in the wastes (for example sewage sludges which turn septic), or due to stockpiling wastes. Research is ongoing to develop ways of predicting odour from new works and levels of complaints from residents. The impact of existing works may be assessed by recording the number of complaints per day over a period and comparing these with activities occurring at the treatment works. Mitigation measures involve first determining where and why odours arise, followed by the implementation of steps to reduce the problem. For example, the design of the treatment works may mean that tanks are difficult to empty thoroughly or that the way the site is operated results in wastes being stored for too long. Consequently, treatment processes or site operations can be modified. Processes that cause odours can be covered, with the air extracted through odour control ventilation systems. Chambers and manholes can be sealed. Some odours may be suppressed by adding chemicals to the waste. It should be noted that older treatment sites were not necessarily designed with odour control in mind, resulting in the need for bolt-on technology and rising costs to mitigate residual odours. However, for new plants much can be done to reduce odour at the site. A slight detectable odour may be noticed near dedicated sewage sludge incinerators, but it is not thought to be a significant impact on local residents.

Box 5.1 Bellozanne sludge dryer, Jersey

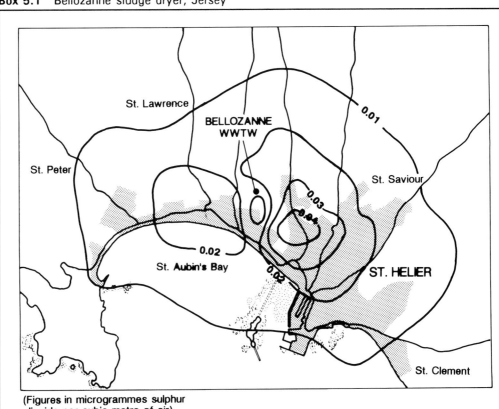

(Figures in microgrammes sulphur dioxide per cubic metre of air)

Assessment of the air quality impact of a proposed sludge dryer at the Bellozanne wastewater treatment works in St Helier, Jersey, was achieved through the characterisation of stack emissions and the prediction of the dispersion of a plume discharged from the stack using an atmospheric pollution mathematical model. The model determined the concentrations of airborne pollutants at various distances from the stack. These predictions were then compared with the relevant air quality standards to assess the effect of the stack on the air quality in the surrounding area and thus allow for appropriate design/mitigation actions to be taken.

Incineration of sludges results in the emission of heat, gases and particulate matter from the stack, and an ash deposit from the incinerator. The quantity and type of pollution from the stack will depend on the initial chemical composition of the sludge, and co-incinerated material, the type of incinerator, its operation and the pollution control measures installed in the stack. The types of pollutants from the stack would include carbon dioxide, nitrogen oxides, heavy metals, dust and low levels of a large number of pollutants. Emissions from the stack may be visible as a plume causing a visual impact, although this can be avoided by raising the temperature of gases in the stack. The case study in Box 5.1 outlines procedures to predict the impacts of sludge dryers and incinerators on air quality.

Emissions to the atmosphere now come under IPC and are controlled by HMIP or local authorities through the issue of licences. To ensure that the emission standards are met, it is important to control the input of wastes into the sewerage system to avoid difficult or dangerous wastes mixing with the sludges. The incinerator needs to be run efficiently so that combustion temperatures and duration are maintained. Finally, pollution control technologies may have to be installed such as an electrostatic precipitator and wet or dry scrubber to remove particulate matter and gases from the stack.

Table 5.9 Examples of the impact of water supply and wastewater treatment works on air quality

Issue	Possible cause	Typical effects	Predictive techniques	Appropriate standards	Mitigation and enhancement options
Water supply					
Dust	Entrainment of dust from stockpiles and haulage roads	Public nuisance	Dust modelling e.g. US EPA air quality models		Dampen roads, cover stockpiles, use wind breaks
Fogs and mists	Increased water vapour in atmosphere	Increase incidence of fogs and mists	Calculate water balance		
Wastewater treatment					
Odour	Processes and operation of works	Public nuisance	Examine histories of complaints and site operation		Design works to avoid odour e.g. cover units, extract air through odour control equipment
Stack emissions	Emissions from incinerators and sludge dryers	Reduce air quality, public health, visual impact of plume	Plume dispersion models	Council Directives on incineration and air quality standards	Control composition of wastes. Install pollution control technology

5.4.2 Water quality

Water supply infrastructure schemes

The quality of surface and subsurface waters may be affected in a number of ways as shown in Table 5.10. Reservoirs may become stores for pollutants from non-point sources, such as nitrates from agricultural land, and from point sources, such as effluent discharges. The increase in nutrients can lead to eutrophication, which seriously affects water quality and the ecology of the lake. Reservoirs may also become infected with micro-organisms, which may pose a health risk to humans if the water is not treated: for example, the water-borne parasite *Cryptosporidium*. Downstream from a reservoir, a reduction in river flow near effluent discharges may lead to increased concentrations of pollutants in the river, and near estuaries the low river flows may result in the upstream penetration of saline water. Inter-basin water transfers may affect water quality in several catchments.

Water quality in reservoirs is difficult to predict. The likely impact of reduced flow on water quality of rivers and estuaries can be modelled given data on the distribution of pollutants in the fluvial system and hydrological parameters. Water quality problems are mitigated by including these issues in the design of reservoir operating rules and restricting effluent discharges near abstraction points.

Water quality problems arising from runoff from agricultural land are being mitigated by the adoption of nitrate sensitive areas (NSA).

Box 5.2 Water quality modelling

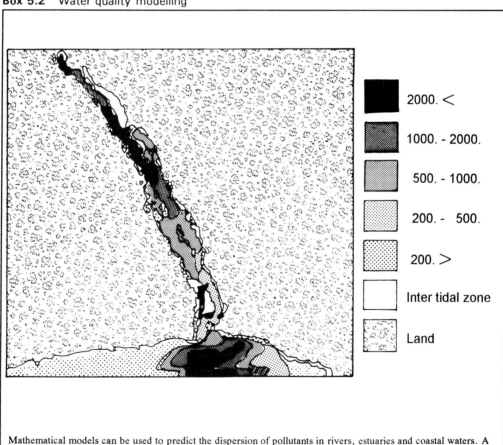

Mathematical models can be used to predict the dispersion of pollutants in rivers, estuaries and coastal waters. A hydrodynamic model is used to derive water depths and current speeds, which are used to simulate pollutant transport and water quality. Water quality determinants may include temperature, for example to model thermal plumes from a power station, or pollutants such as faecal coliforms, ammonia and chlorine to model effluent discharge from an outfall. The model results need to be calibrated and verified against field data, then the impact of a scheme on water quality can be simulated for different conditions. The significance of the impact is determined by comparing the model output with water quality standards.

Table 5.10 Examples of the impact of water supply and wastewater treatment works on water quality and water resources

Issue	Possible cause	Typical effects	Predictive techniques	Appropriate standards	Mitigation and enhancement options
Water supply					
Waterlogging in valley	Reservoir	Local rise in water table	Hydrogeological modelling		Control level in reservoir; grouting; reservoir bed lining
Change fluvial regime	Regulation of river	Dampen variations in discharge; maintain above-average flows during droughts; lower flows during rainy periods	Hydrological modelling		Develop operating rules for reservoir
Water quality in reservoir	Accumulation of nutrients and algal growth in reservoir	Eutrophication in reservoir; pollute groundwater and downstream waters	Water quality modelling	UK regulations for nitrate sensitive areas and water quality standards for surface waters	Reduction in fertiliser applications; appropriate levels of water treatment; destratification
Downstream water quality	Lower flows	Higher concentrations of pollutants downstream from effluent discharge; upstream intrusion of brackish water	Water quality modelling	EC Directives on water quality	Determine acceptable minimum flow regime
Lower groundwater levels	Over-pumping	Loss of springs; low river flows; loss of wetland areas; declining fish numbers	Hydrogeological modelling		Limit or redistribute abstractions; reduce leakage; augment river flow from alternative sources
Wastewater treatment					
Pollution from effluent discharge	Unauthorised pollutants in trade effluent; insufficient or inappropriate treatment; discharge of effluent to sensitive waters; discharge of treatment chemicals eg chlorine	Poison and kill wildlife; odours; scum; health risk to bathers	Hydrodynamic and water quality modelling	Urban waste water directive	Strict supervision of discharge consents; level of treatment appropriate for receiving waters
Sludge disposal	Runoff from agricultural land; leachate from landfill	Lower water quality; affects ecology	Model runoff and leachate generation	EC Directives on sludge disposal to agricultural land; IPC	Use appropriate waste management procedures

Wastewater treatment works

Effluent discharges may lower the quality of the receiving waters, affecting water chemistry, microbiology, water temperature, turbidity and colour among other things. Considerable studies have been undertaken on long sea outfalls leading to the establishment of design guidelines for environmental protection (Neville-Jones and Dorling, 1986). The impact of discharges on water quality can be modelled as described in Box 5.2. The significance of the impact will depend on the concentration of pollutants in the discharge and the state of the receiving waters. Small streams, sites inhabited by uncommon flora and fauna, and important fisheries may be highly sensitive to water quality. Significance is determined by comparison of predicted water quality and the EQSs for the receiving waters. Water quality problems from outfalls are mitigated through the control of emissions by the NRA and HMIP.

5.4.3 Water resources

Water supply infrastructure schemes

The impoundment of rivers and abstractions from rivers and groundwater have increasingly greater impacts on water resources as attempts are made to ensure that supply meets demand (Table 5.10). River impoundment has an immediate effect on downstream river regimes, with a decrease in high flows and regulation of the river regime in order to meet diverse objectives such as water supply to downstream abstraction points, improve base flows for ecological or water quality purposes and to attenuate floods. Abstractions from groundwater and rivers reduce water stored in these systems, which can lead to the depletion of aquifers, the loss of base flow in rivers and the loss of wetlands. Box 5.3 illustrates the widely documented reduction of low flows in streams which in the past have depended on base flow from the aquifer. Finally, the redistribution of waters within and between catchments by pumping water from one location to another is aimed at meeting supply in one area by taking surplus water from another, but may result in large-scale environmental impacts. The Roadford reservoir scheme illustrates the scale of such projects and the type of environmental studies required to mitigate impacts (Box 5.4).

Box 5.3 Thames chalk streams low flow study

Six chalk streams were found to be suffering from low flows caused by groundwater abstraction. The River Darent, once one of the finest trout streams in England, had sometimes dried up completely and stretches of the River Ver and River Misbourne were also gradually drying out, to the detriment of wildlife and the amenity of many villages along their lengths.

Once the cause of the problem had been established, remedies could be identified. In most cases, low summer flows could be increased by pumping water from new boreholes located well away from the rivers. The improvements in flows could be accompanied by sealing the river beds to prevent leakage by lining them with modern geotextiles or using the ancient method, puddled pulverised chalk. The creation of pools and riffles and replanting chalk stream vegetation could provide new habitats for trout and other wildlife.

There are standard hydrological methods available for determining the water resources of a region. In order to develop water supply, it is necessary to compare demand forecasts with an analysis of existing water resources. If a supply deficit is projected, it could be reduced by improving the efficiency of the existing supply system, by increasing water storage or by increasing abstractions. However, given the predetermined climatic and geological conditions, water resources are finite, so the mitigation of the effects of over-exploitation of the water resources needs to start to look at ways of reducing demand as well as maximising existing resources.

Wastewater treatment works

Effluent discharge from a wastewater treatment works may affect water resources locally if the quantities are large relative to the receiving waters.

Box 5.4   Roadford reservoir scheme

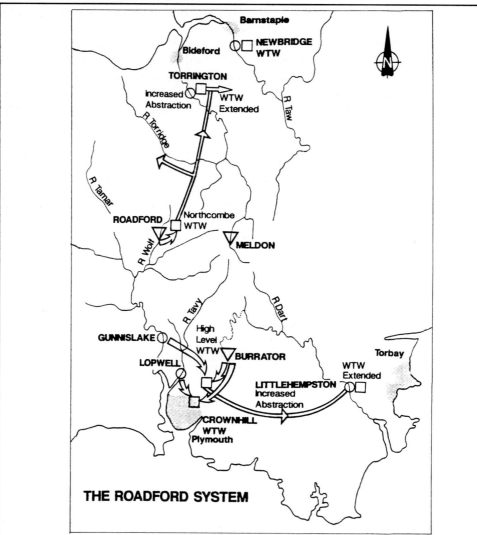

The Roadford reservoir scheme, shown here schematically, is a complex water resource scheme incorporating virtually all types of water resources engineering development: impoundment, river intake works, transfer pipelines and water treatment works. The reservoir is designed to operate in conjunction with abstractions from the Rivers Tamar, Tavy, Dart, Torridge and Taw, as well as the Burrator and Meldon Reservoirs. The concept of the scheme is that abstraction from the various rivers is increased from present levels at times when there is sufficient water in the rivers; during dry periods, the river abstraction is reduced and the increased supplies are obtained instead from Roadford. There is therefore considerable potential for environmental impact, both detrimental and beneficial, not just in the immediate vicinity of the new reservoir but on all of the rivers and reservoirs included within the scheme. To manage the scheme effectively an integrated system of control rules has been proposed to minimise detrimental environmental impacts while maintaining scheme yield (Lawson *et al.*, 1991).

5.4.4 Land

Water supply infrastructure schemes

The creation of reservoirs causes the loss of land resources, including agricultural land, ecological sites, open space of amenity value and mineral resources. It may also affect slope stability near the reservoir, seismic activity and soil erosion rates on recently exposed excavations (Table 5.11). The installation of pipelines and the creation of borrow pits to provide aggregate also affect soils, the landscape and ecology. Any impacts on land need to be assessed by thorough site investigations prior to site selection. Avoiding sites with valuable resources and geotechnically unsuitable ground conditions is the major means of mitigation. The adoption of alternative options such as improving leakage control would also reduce impacts.

Wastewater treatment works

The main effect of wastewater treatment works on land arises through potential contamination of soils with wastes (Table 5.11). This may include dangerous substances and pathogens in sludges applied to land or stored at the works site. The application of sludges to agricultural land is controlled through UK legislation and Council Directives. This aims to avoid the dangerous build-up of heavy metals in soils and existence of pathogens in the soil by setting treatment standards, application rates, limits on heavy metal concentrations in sludges, recording and checking systems and agricultural practices regarding means of application and harvesting procedures.

Table 5.11 Examples of the impact of water supply and wastewater treatment works on land

Issue	Possible cause	Typical effects	Predictive techniques	Appropriate standards	Mitigation and enhancement options
Water supply					
Sterilisation of mineral resources	Permanent inundation; building over mineral resources	Sterilise sand and gravel deposits	Examine mineral resource plans; geological maps of surface materials	Consult with County Council	Avoid sites with valuable mineral reserves; exploit prior to inundation
Slope stability	Damming leads to saturation of lower hillslopes, high pore water pressures and unstable slopes	Slope failures around dam; increased sedimentation and turbidity; affect stability of dam, amenity of lake side and ecology	Geotechnical investigation	Geotechnical procedures	Thorough site investigation and design
Seismology	Inundation increases pressure on land surface above faults; lubrication of faults by seepage	Increase incidence and potency of earth tremors	Geological investigation	Design standards for dam	Avoid tectonically unstable areas; design dams following standards for seismology
Soil erosion	Heavy rain during excavation cuttings and earthworks	Loss of soil making vegetation recovery difficult later; high suspended sediment loads in river affecting fish; high deposition rates affecting channel morphology	Estimate soil erosion rates		Reduce areas of bare soil, protect from rain using geotextiles; undertake earthworks during good weather; replant as soon as possible; runoff from sites channelled to settlement lagoons
Wastewater treatment					
Soil contamination	Application of sludges to land	Build-up of heavy metals in the soils; pathogens in untreated sludges affecting cattle	Study soil chemistry and application rates	Sludge (Use in Agriculture) Regulations 1989	Follow agricultural practices set out in legislation; control emissions to treatment works to improve quality of sludges

5.4.5 Physical processes

Water supply infrastructure schemes

The regulation of rivers by dams has a well-recorded impact on the geomorphological processes of erosion, sediment transport and deposition (Table 5.12). Upstream of the reservoir the decrease in velocity of the river leads to deposition and the build-up of a lag deposit which merges into a delta. The decrease in channel slope over time tends to encourage an increase in channel sinuosity, and the low-lying area around the dam becomes increasingly prone to flooding. Downstream of the dam the river is normally reduced in size and its river channel morphology is no longer in equilibrium with the channel-forming processes. Sometimes the reduced volume of the flow leads to a significant reduction in stream power and there is little further change in bed morphology. However, during high flow, the release of clear water from the reservoir can cause severe channel erosion. These issues may be assessed through geomorphological process studies and modelling.

Wastewater treatment works

Scour around outfalls is the only significant potential impact of wastewater treatment works on physical processes. The potential scour could be modelled, although it would be more likely just to revet the banks around a river outfall and protect sea outfalls.

Table 5.12 Examples of the impact of water supply and wastewater treatment works on physical processes

Issue	Possible cause	Typical effects	Predictive techniques	Appropriate standards	Mitigation and enhancement options
Water supply					
Channel change	Change in fluvial regime following dam construction	Accelerated rates of deposition leading to meandering upstream of dam; accelerated erosion during clear water releases from dam downstream	Geomorphological process studies		
Siltation	Sedimentation in reservoir	Long-term reduction in storage	Geomorphological studies and modelling sedimentation patterns		Check dams in tributaries; periodic dredging
Wastewater treatment					
Scour around outfall	Erosion of banks or rock armour	Bank instability	Model scour		Locate outfall to minimise scour; protect banks

5.4.6 Impact on natural habitats

The main types of impact on natural habitats involve:

- physical damage during construction or operation
- the permanent loss of habitats and species
- the modification of habitats
- the creation of new habitats.

These are summarised in Table 5.13.

Water supply infrastructure schemes

New reservoirs can have considerable impacts on ecology in a catchment (Table 5.13). Inundation results in the permanent loss of habitats, and the operation of reservoir releases can modify downstream habitats. The reservoir site may experience increases in animal populations: for example, midges and birds. The latter has been known to have a bearing on the operational safety of aircraft approaching or leaving nearby airports. Sometimes the regulation of rivers leads to less habitat diversity, which favours some species at the expense of others. As a consequence total biomass may be maintained but the diversity or abundance of species changes. The loss of spate flows can have detrimental effects on the movement of migratory fish as described in Box 5.5. A reduction in flooding events may lead to changes in the ecology of flood meadows and other wetlands.

Rivers flowing over different substrates have distinct differences in water chemistry. They also support different assemblages of flora and fauna; indeed a methodology has been developed to classify rivers according to their macrophytic flora (Holmes, 1984). Changes in water quality arising from the mixing of waters from different catchments from inter-basin water transfers can potentially modify the river corridor ecology.

An evaluation of the impacts of water supply infrastructure schemes requires:

- baseline surveys and data collection to determine the quality of the site
- hydrological studies to predict flows
- a qualitative or, where possible, quantitative assessment of the impact of flows on ecology.

Standard procedures have been developed to undertake ecological surveys of river corridors (for example, Phase I and Phase II ecological surveys, and macrophyte flora surveys). The quality of these sites in ecological terms may depend on several factors. English Nature has proposed criteria for categorising a site based on its size, rarity, fragility, naturalness, diversity, typicalness, record history, intrinsic appeal and potential value. The surveying and evaluation of these sites need to be undertaken by suitably qualified staff. The impact of the scheme on ecology may be based on a qualitative assessment or the use of programs such as PHABSIM or RIVPACS, which correlate environmental and ecological parameters.

Mitigating the negative impacts of water supply schemes involve avoiding ecologically valuable sites, water resources management and creating new habitats. The latter can be difficult at reservoirs owing to fluctuations in the water level, and needs to be designed carefully.

Table 5.13 Examples of the impact of water supply and wastewater treatment works on natural habitats

Issue	Possible cause	Typical effects	Predictive Techniques	Appropriate standards	Mitigation and enhancement options
Water supply					
Permanent inundation	Reservoir	Loss of habitats; eutrophication due to decomposition of inundated vegetation			Select sites of low ecological importance; clear excessive vegetation; fell timber
Changes to river corridor ecology	Unsympathetic reservoir operation; excessive abstraction	Changes in species number and diversity	Model operating rules and compare discharges with ecological requirements		Set minimum flow
Impede fish migration (especially salmon and sea trout)	Lack of flow or summer spates; physical barriers	Fish do not enter river, or limit their movements upstream	Model summer flows and compare with data on fish movements		Develop spate sparing rules, maintain flows and water quality
New habitats	Creation of water bodies	Attract wildlife; optimise fisheries in new lake	Ecological studies		Develop habitats by water's edge to encourage wildlife
Degradation of wetland habitats	Reduction in frequency and magnitude of flooding	Loss of flora and fauna	Hydrological modelling and ecological observation		Develop operating rules to maintain sufficient flows
Impact on ecology of estuaries	Regulation of river flow; changes in river quality	Change in distribution of marine, estuarine and freshwater species; changes in bird population following changes in food source	Water quality and hydrological modelling coupled with ecological studies		Maintain minimum flows
Wastewater treatment					
Changes in aquatic ecology around outfall	Settlement of solids on benthic organisms; relatively higher nutrient levels in water	Lower numbers of species around outfall, and possible increase in number of individuals of certain species	Examine case histories	EC Directive on shellfisheries 78/639/EEC Design standards for outfalls	Upgrade level of treatment; accept limited degradation
Health hazard to animals and humans	Spread of untreated sludge on agricultural land	Endanger health of grazing animals and humans consuming produce		EC Directive and UK Regulations on sludge applied to agricultural land	Follow guidelines on methods of treatment and application; soil injection techniques
Water quality impact on ecology	Discharge of toxic wastes; chemicals used in treatment e.g. chlorine and lime, excessive phosphate	Poison flora and fauna; cause eutrophication in water bodies	Case studies, model water quality and impact on flora and fauna eg RIVPACS	EQS and EQO of receiving waters	Improve treatment

Box 5.5 Environmentally sensitive operating rules for the Roadford scheme

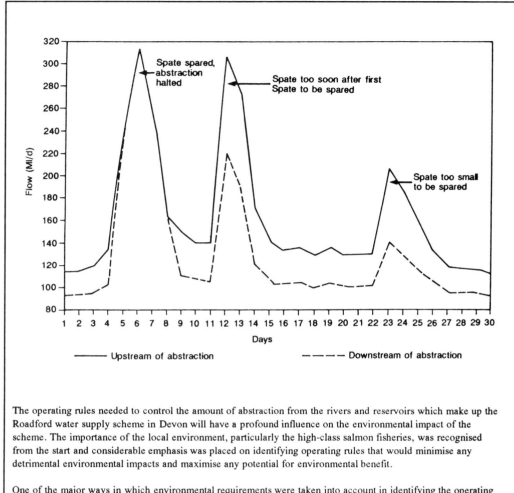

The operating rules needed to control the amount of abstraction from the rivers and reservoirs which make up the Roadford water supply scheme in Devon will have a profound influence on the environmental impact of the scheme. The importance of the local environment, particularly the high-class salmon fisheries, was recognised from the start and considerable emphasis was placed on identifying operating rules that would minimise any detrimental environmental impacts and maximise any potential for environmental benefit.

One of the major ways in which environmental requirements were taken into account in identifying the operating rules was the development of the concept of spate sparing into a more formalised operating rule. The detailed requirements for spate sparing will vary between rivers. However, the principle remains the same. The movement of migratory fish upstream has been observed to be related to spate flows, particularly those that occur after a period of low flows. Therefore, if abstraction can be halted, or reduced, for a short time during those critical spates it could be possible to reduce significantly detrimental impacts on fish migration without leading to a significant reduction in scheme yield.

Wastewater treatment works

The main impacts on ecology from wastewater treatment works arise from effluent discharge in aquatic environments. Other impacts include the loss of habitat due to the construction of a new site and effects on flora and fauna following the use of sludges in forestry or for land reclamation (Table 5.13).

Effluent discharges may affect aquatic ecology owing to relative differences in temperature, toxic chemicals, excess nutrients and suspended solids. Water used by the power generator companies for cooling is usually warmer than the receiving waters and may form a volume of water used preferentially by some species of fish. It may also contain chemicals such as chlorine to keep inlet pipes free from organisms. Accidental spillages of toxic chemicals may lead to fish kills while effluent discharges may result in long-term impoverishment of aquatic ecology. High inputs of nutrients to small water bodies with little mixing may cause eutrophication. Finally, solids in discharges, from sewage for example, may block fish gills and smother benthic organisms (see also Chapter 7).

Measures to identify, evaluate and mitigate the impacts of sea outfalls on flora and fauna have been well established (Neville-Jones and Dorling, 1986; Dorling, 1988; ICE, 1989). These involve:

- baseline surveys
- assessing water quality output models
- toxicity experiments on fauna near outfalls e.g. shellfish
- controlling effluent quality.

The impact of toxic effluent on ecology has to be mitigated through emissions control.

The construction of wastewater treatment works and outfalls may result in localised losses of, or impoverishment of, habitats. Consequently, important ecological sites and fisheries should be avoided altogether. On land, planting schemes may help to mitigate vegetation losses although rapid planting solutions may not always provide the best long-term ecological mitigation measures.

5.4.7 Community structure

The development of water supply projects involving the creation of reservoirs may have very severe impacts on the local population, depending on the number of communities at risk to flooding. Dams built in highland areas are less likely to affect communities than those in lowland valleys as the latter are more likely to be populated. However, isolated rural communities may be significantly disrupted. If a new reservoir results in drowning properties, the inhabitants need to be relocated. This may affect not only the individuals concerned, but also the receiving community. The main mechanism to mitigate this effect is to avoid choosing sites containing residential properties. If this cannot be avoided, then studies of the attitudes and aspirations of the local residents, together with consultation and the provision of suitable housing, is required.

5.4.8 Health and safety

Water supply infrastructure projects may affect the health and safety of the population through the supply of contaminated water (Table 5.14). There have been several recent cases of accidental chemical spillages into the water supply system and the presence of microbes in drinking water that cause illness. These problems can be resolved by adopting suitable treatment methods, monitoring water quality and meeting EQSs.

Events such as dam failure are rare but have catastrophic impacts on health and safety. This is avoided through design to engineering standards and regular inspection.

Emissions to water, air and land as a result of wastewater treatment pose several environmental health issues. There has been concern over the health of bathers (Jones et al., 1990), people undertaking water contact sports and eating fish and shellfish caught near sewage outfalls. Incinerators in general have been implicated in health studies. Sludges applied to land may contain pathogens and heavy metals, which pose short- and long-term health risks. These issues can be mitigated by emission controls, meeting EQSs and following guidelines on sludge application to land.

Vermin may be a problem at wastewater treatment works (especially caused by screenings disposal practice) and there is a very small risk that this would pose a health risk to humans. Good housekeeping can reduce the numbers. Measures such as good site security, supervision of the works and maintenance reduce the likelihood of accidents.

Table 5.14 Examples of the impact of water supply and wastewater treatment works on health and safety characteristics

Issue	Possible cause	Typical effects	Predictive techniques	Appropriate standards	Mitigation and enhancement options
Dam instability	Slope instability; seismic tremors; poor construction or design	Cracks in dam; failure; loss of life; inundation over large area	Geotechnical modelling; structures modelling	Engineering design standards	Initial site investigation; good design and construction; inspection; contingency plans
Community health	Consumption of contaminated shellfish, or fish; respiration of air pollutants; microbiological quality of drinking water	Short- and long-term illness	Air pollution and contaminant dispersion modelling; analysis of water and air samples	EC Directives on water and air quality; local planning authority; Health and Safety Executive	Maintain EQS; educate people on cooking methods
Health of construction and maintenance staff	Pathogens in existing sewerage systems; fumes; odours	Illness		Health and Safety Executive; local planning authority	Follow health and safety procedures
Land contamination	Sludge disposal to land	Build-up of heavy metals; presence of pathogens		EC Directives and UK regulations	Follow guidelines on use of sludge in agriculture
Accidents and deliberate spillages	Slurry spills off agricultural land; unauthorised tipping into water supply; breaches of consent licences for discharges to public sewers	Water contamination			Inspect water supplies; regular analysis of water samples

Table 5.15 Examples of the impact of water supply and wastewater treatment works on socio-economic parameters

Issue	Possible cause	Typical effects	Predictive techniques	Appropriate standards	Mitigation and enhancement options
Water supply					
Resettlement	Inundation for reservoir	Relocation of inhabitants of individual properties, hamlets and villages	Compare new water levels with maps of settlements		Avoid populated sites; relocate people to similar or better housing
Land values	Depreciation e.g. due to restriction on activities around reservoirs or increase e.g. due to leisure developments	Change in land values around reservoir	Economic studies		Develop infrastructure for amenity development
Wastewater treatment					
Land values	Siting new works	Land and house values depreciate as occupiers dislike odours and visual impact of works			Good site selection; landscaping and screening

5.4.9 Socio-economics

Table 5.15 sets out typical socio-economic impacts. The construction of dams and treatment works generally has a low impact on the local socio-economics of an area. Construction may lead to only relatively a low amount of temporary employment, and also requirements for raw materials which may be available in the area, and labour and material requirements remain low following construction. Reservoirs may attract local business as a result of the development of leisure activities outlined below. Wastewater treatment works would have relatively little impact on the socio-economic character of an area, although land and property values could be affected.

5.4.10 Leisure and amenity

New reservoirs built to ensure water supply are often seen as opportunities to develop recreational facilities including fishing, water contact sports, and rambling (Table 5.16). The types of activities may be affected by the changes in water level in the reservoir so that access needs to be provided to the water's edge during drawdown when extensive muddy banks may be exposed. Studies of the contours of the proposed reservoir will indicate the extent of exposed shoreline during drawdown, which can be used as a basis for designing recreational facilities. Consultations are also required among hydrologists, conservationists, landscape architects and developers in order to design a multi-use strategy for the reservoir.

The construction of wastewater treatment works may have a negative effect on the leisure and amenity value of an area arising from perceived or actual changes in environmental quality. This is particularly important for outfalls located near waters with a recreational value, for example fisheries and bathing waters. This effect should be minimised by siting the works and outfalls at acceptable distances from popular and designated beaches and meeting EQOs for the use of these waters.

5.4.11 Land use and landscape

The creation of a reservoir has a major impact on land use and landscape (Table 5.17). These impacts are assessed by sketching views of the proposed scheme with different water levels. Mitigation measures consist of designing planting schemes to reduce or enhance the visual impact. These should involve habitat creation for nature conservation.

Wastewater treatment works can have a significant visual impact, particularly if they include tall buildings and a chimney stack for an incinerator. The visual impact can be mitigated by creating earth bunds and planting around the site, designing tall buildings carefully and building structures underground.

In both cases good public relations, established from an early stage of the project and maintained throughout the planning and construction phases, is an essential element in mitigating the impacts on land use and people.

Table 5.16 Examples of the impact of wastewater supply and wastewater treatment works on leisure and amenity

Issue	Possible cause	Typical effects	Predictive techniques	Appropriate standards	Mitigation and enhancement options
Water supply					
Reduced access	Fence off construction site, site works	Limit access across countryside or along beach		Footpath Order	Signpost information on works and diversion routes
Impact of construction traffic on local inhabitants	Movement of staff and materials to and from site	Traffic, delays, pollution from exhaust emissions	Estimate number of vehicles at stages of project	Agree access routes with local authority	Programme works; arrange diversions
Visitor attractions	Site works				Safety measures
New development opportunities	Open water bodies	Opportunities for sailing, water contact sports, angling and walking			Landscaping; habitat creation; provision of services, car parks, picnic areas, footpaths etc.; develop operating rules to optimise use of reservoir for water supply, ecological and amenity uses
Inhibit water transport	Construction of dam	Inhibit rowing, and sailing along the river			Provide access around the dam
Wastewater treatment					
Reduce amenity value	Unsightly location; odour; worries of health risks bathing or fishing near outfalls; loss of sites of nature conservation value	Reduce number of visitors			Site works away from amenity areas; use odour control measures; landscaping; appropriate level of treatment of discharge; habitat creation
Noise	Pumping	Reduce number of visitors			Soundproof buildings, use less noisy equipment

Table 5.17 Examples of the impact of water supply and wastewater treatment works on land use, landscape, and cultural heritage

Issue	Possible cause	Typical effects	Predictive techniques	Appropriate standards	Mitigation and enhancement options
Temporary visual intrusion	Construction site; removal of vegetation	Intrusion of views in open countryside	Landscape assessment	Local planning authority; Countryside Commission	Minimise area affected; screen site; reinstate after use
Permanent visual intrusion	Large structures such as dams, chimney stacks and buildings	Visual intrusion and obstruction	Landscape assessment	Local planning authority; Countryside Commission	Design structures; site location; screening by vegetation or building; siting small buildings in hollows or lower ground
Change in land use	Reservoir construction	Loss of agricultural land	Field surveys		Financial compensation
Loss of heritage sites	Inundation or damage during construction	Inundation of sites of historical or cultural significance and nature conservation	Examine new water levels and maps of heritage sites	Countryside Commission, English Heritage	Carry out archaeological surveys and exploratory digs; avoid sites; make permanent records; relocate sites; habitat creation

5.4.12 Noise

Noise can be an issue during construction and operation of plant, and depends to an extent on the proximity of the site to residential areas. BS 5228 can be used to estimate noise levels during construction, and a number of measures are available to reduce construction noise, particularly acoustic screening. The local planning authorities often set restrictions on noise levels, which have to be adhered to during construction.

During operation, plant such as pumping stations and extractor fans emit noise, which can be mitigated by acoustic screening within the building and maintaining plant. With appropriate building design, noise is not expected to be a significant problem at wastewater treatment works and sludge incinerators. However, in rural areas with low levels of background noise, the operation of wastewater treatment works has given rise to objections from local inhabitants on noise grounds.

5.4.13 Traffic and navigation

Construction creates a certain amount of traffic by transporting raw materials, waste and the workforce. At a dam site the numbers of vehicles may be considerable owing to the large quantities of raw materials required. Construction may also impede existing traffic on the roads by requiring temporary lane closures (for example due to pipe laying), reducing car parking facilities (by using open space for a construction site) or increasing journey delays (by the movement of slow-moving vehicles).

The impact of traffic during construction needs to be assessed by attempting to predict the likely number of vehicles per day during typical stages of the project. This figure is then compared with traffic counts on the roads most likely to be used by construction traffic.

River regulation following reservoir construction may significantly affect navigation downstream. It would be necessary to maintain minimum flows to allow boat passage at all times.

During the operation of wastewater treatment works tankers would be required to transport sludges to disposal sites. Transporting sludges by tanker from rural sites can cause congestion problems on narrow roads. The significance of this operation would be assessed as above.

5.4.14 Cultural heritage

The cultural heritage of an area includes sites such as scheduled ancient monuments, non-scheduled archaeological sites, listed buildings and conservation areas, but also historic landscapes, parklands and gardens. The construction of a new dam can cause a very great impact on the cultural heritage by virtue of the large area affected by inundation. Once archaeological investigation is focused on an area it is not uncommon for a large number of new sites to be discovered, reflecting the lack of systematic archaeological investigations across the country. Even comparatively smaller sites such as wastewater treatment works may have an impact on archaeological sites as many urban and rural areas have been occupied for a long period.

5.5 CONCLUSIONS

The need for EAs was adopted in European legislation in 1985 and given legal backing in the UK in 1988. For some projects a form of EA has been in existence for a much longer period; however, widespread adoption of EA has occurred since the late 1980s. As a result, methods for undertaking an EA are still evolving, and there are many areas required for research.

Legislation has not only opened the way for undertaking EAs for a range of projects, but has also led to increasing numbers of construction projects. For example, the water service companies are undertaking large capital works programmes to improve drinking water and wastewater treatment. IPC will probably lead to capital investment in effluent treatment at many industrial premises. In some cases the most economic, although not necessarily the most effective, measures involve bolt-on technology, whereas at green field sites there are greater opportunities for designing pollution control and environmental mitigation measures. The likely result of this is an increase in demand for EA.

5.6 REFERENCES

DoE (1992)
Digest of Environmental Protection and Water Statistics 1991
HMSO No. 14, 103 pp

DoE/Welsh Office (1989)
Environmental Assessment: A guide to the procedures
HMSO, 64 pp

DORLING, C. (1988)
Site investigation for outfall design, construction and monitoring. A discussion document
WRc Report No. ER315E

FROST, R., POWLESLAND, C., HALL., J.E., NIXON, S.C. and YOUNG, C.P. (1990)
Review of sludge treatment and disposal techniques
HMIP Report No. PRD 2306-M/1, 537 pp

HOLMES, N. (1984)
Typing British Rivers According to their Flora
Nature Conservancy Council (English Nature), Peterborough

ICE (1989)
Long Sea Outfalls
Thomas Telford (London), 230 pp

IPHE (1986)
Construction and Maintenance of Long Sea Outfalls
Technical and Training Symposium, 16-17 September 1986
Institute of Public Health Engineers (IWEM), London

JONES, F., KAY, D., STANWELL-SMITH, R. and WYER, M. (1990)
An appraisal of the potential health impacts of sewage disposal to UK coastal waters
Journal Institution of Water and Environmental Management, London, Vol 4, pp 295-303

LAWSON, J.D., SAMBROOK, M.T., SOLOMON, D.J. and WEILDING, G. (1991)
The Roadford scheme: minimising environmental impact on affected catchments
Journal Institution of Water and Environmental Management London, Vol 5, pp 671-681

NEVILLE-JONES, P.J.D. and DORLING, C. (1986)
Outfall design guide for environmental protection
WRc Report No. ER209E, Swindon

POWLESLAND, C. and FROST, R. (1990)
A methodology for undertaking BPEO studies of sewage sludge treatment and disposal
WRc Report No. PRD 2305-M/1, 195 pp, Swindon

SMY, E.D.A. (1989)
Construction of pipeline outfalls
In *Long Sea Outfalls*, Ed ICE, pp 117-130
Thomas Telford

CIRIA Research Project 424

Environmental Assessment

A guide to the identification and mitigation of environmental issues in construction schemes

6. Linear development

Chris Ferrary, Technical Director, Environmental Resources Management

6 Linear development

6.1 TYPE OF SCHEME CONSIDERED

This chapter is concerned with the environmental impacts of linear engineering projects. This essentially means transport infrastructure projects such as highways and railways (including light rail schemes), but is also taken to include pipelines for the transport of oil, gas or other such commodities and transmission lines for electricity.

The various types of development examined in this chapter are listed in Table 6.1. The environmental effects associated with the construction of linear developments for transport infrastructure can be significant merely because they may extend for substantial distances. In rural areas, topography and natural features may dictate route alignments and pose specific problems, while in urban areas, complex land-use patterns and the need to serve the transport needs of the community mean that proposed routes often run through busy commercial, industrial and residential areas. Consequently, construction often has to take place in highly constrained situations with potential disruption to traffic and communities alike. Similarly, during operation, issues of pollution, noise, vibration and visual impacts (amongst other effects) are likely to arise.

Table 6.1 Examples of linear development projects

Scheme type	Typical promoter	Type of works
Highways	Department of Transport	Motorways Trunk roads
	Local authority	Urban and rural road schemes
Railways	British Rail	Conventional rail
	London Underground Ltd	Underground/Metro
	Passenger transport executive/local authority	Light rail
Pipelines	Statutory undertakers/private companies	Gas main Petroleum pipelines
Transmission lines	Statutory undertakers/private companies	Electricity supply

6.2 KEY STANDARDS, LEGISLATION AND ADVICE

6.2.1 Introduction

Transport and the provision of transport infrastructure are central to the environmental debate. They raise key issues about mobility, urban congestion, use of resources and pollution. In terms of considering these issues, some, such as the contribution of carbon dioxide emissions from vehicles to the greenhouse effect, are only appropriate to be considered at a strategic level. However, the potentially significant effects on the environment that may arise from individual infrastructure projects often mean that these need to be considered in an Environmental Assessment (EA). While EA is a formal procedure, which is a statutory requirement in relation to some forms of specified linear development (see Section 6.2.2), the consideration of environmental issues is often also important in the planning of more minor schemes, such as estate roads. It should also be noted that while the general discussion of EA for linear development in this chapter inevitably concentrates on the procedural aspects, in practice the process is often an iterative one where the consideration of environmental issues can inform the design of the scheme.

6.2.2 Requirements for EA

As with EA for all development projects, the primary legislation that places a requirement on developers of transport infrastructure to undertake an assessment of the effects on the environment which this may have is the European Community (EC) Directive 85/337. In the UK, this has been incorporated into UK law through the enactment of sets of regulations which have been issued under the provisions of various existing Acts of Parliament.

Different regulations apply for the various types of linear development under consideration in this chapter. The legislation listed below is that which applies generally in England and Wales. Other regulations apply to similar developments to be undertaken in Scotland or Northern Ireland, but their provisions are broadly the same:

- The Town and Country Planning (Assessment of Environmental Effects) Regulations 1988 (SI: 1199) relate to applications for planning permission, including local authority highways works, railways works and pipelines.

- The Highways (Assessment of Environmental Effects) Regulations 1988 (SI: 1241) relate to orders issued under the Highways Acts for highways, trunk roads and motorways.

- The Electricity and Pipeline Works (Assessment of Environmental Effects) Regulations 1990 (SI: 442), for oil and gas pipelines and transmission lines subject to authorisation under the Pipelines Act 1964 or the Electricity Act 1989 by the Secretary of State for Energy.

- Linear developments for which consent is sought from Parliament through a Private or Hybrid Parliamentary Bill (either specifically or as ancillary to a larger development proposal) require EA by virtue of Parliamentary Standing Orders applied to Private Bill procedures (Standing Order 27A).

- Linear developments (e.g. light rail) for which permission is sought under the provisions of the Transport and Works Act 1992.

In describing the provisions of these various regulations, reference is made specifically to SI: 1199 (referred to hereafter as 'the Regulations') as these most fully interpret the EC Directive and integrate it into the UK planning system, and are essentially duplicated by other UK legislation.

The types of development to which the Regulations apply are listed in two schedules attached to the Regulations. EA must be undertaken if developments fall within categories listed in Schedule 1 to the Regulations. This includes 'special roads', i.e. roads where use is restricted to certain types of traffic. The most common type of special road is a motorway, where road users such as cyclists, learner drivers, motorcycles under a certain size etc. are prohibited. Article 7 also includes lines for 'for long-distance railway traffic'.

The EA of projects listed in Schedule 2 is not always mandatory, but must be undertaken if the project is likely to have significant effects on the environment. 3(b) specifies 'transmission of electrical energy by overhead cables'. 10(d) of Schedule 2 includes 'the construction of a road ... not being development falling with Schedule 1'. In principle, therefore, all highway proposals require EA to be undertaken if significant environmental effects are likely. Also 10(g) relates to 'a tramway, elevated or underground railway, suspended line or similar line, exclusively or mainly for passenger transport', while 10(h) specifies 'an oil or gas pipeline installation'.

Section 15 of the Planning and Compensation Act 1991 also allows the Secretary of State for the Environment to make further regulations governing the need for EA.

6.2.3 Government advice on EA

UK Government advice on the application and scope of EA in relation to linear development proposals comes from a number of sources. These are:

- Department of the Environment (DoE) Circular 15/88 (Welsh Office Circular 23/88) *Environmental Assessment*

- DoE/Welsh Office (1989) *Environmental Assessment – A guide to the procedures* (also known as 'the Blue Book')

- Department of Transport (1983) *Manual of Environmental Appraisal* (MEA)

- Department of Transport (1989) *Environmental Assessment under EC Directive 85/337* Departmental Standard HD 18/88.

The relevance of each of these is discussed in turn.

Circular 15/88 is the Government's formal advice on interpreting the Regulations generally, and is specifically aimed at providing guidance for the local authorities whose job it is to administer the EA system. However, it also gives helpful advice to developers whose duty it is to actually carry out the preparation of the Environmental Statement (ES). Among much other guidance, the Circular sets out criteria for deciding whether a highway proposal (other than motorways where EA is mandatory) falls within the meaning of Schedule 2 to the Regulations. Paragraphs 19–20 of Appendix A to the Circular state that an EA will be required for local authority or private road proposals where:

a) In rural areas:

- New roads or improvements are longer than 10 km
- Schemes are longer than 1 km which pass through a National Park or within 100 m of a Site of Special Scientific Interest, or National Nature Reserve or a Conservation Area.

b) In urban areas:

- New roads or major improvements where more than 1500 dwellings lie within 100 m of the centre line.

The 'Blue Book' re-iterates this advice for road schemes, and also points out that motorways and trunk roads not subject to planning regulations are also subject to the same criteria under the provisions of SI: 1241. In relation to pipelines, the 'Blue Book' notes that where these are over 10 miles in length, EA is required if the Secretary of State for Energy considers significant environmental effects are likely.

The MEA sets out the form and content of EA as specifically applied to trunk road schemes and motorways. However, the guidance it contains is also very widely used in practice to assess other types of road schemes (e.g. proposed by local authorities). It has also been used, in a modified form, and applied to the assessment of other types of linear development, particularly railway projects. The MEA recommends the use of a framework approach to provide a summary of the environmental effects of a highway scheme. A number of specific sections also set out methodologies to measure impacts and criteria for their evaluation. In relation to the practice of EA generally, the MEA has an important role as one of the few documents that predates the introduction of mandatory requirements for EA. However, because of this it should be noted that carrying out assessments in accordance with MEA advice does not guarantee that the full requirements of the EC Directive or the subsequent UK Regulations will be met. Although at the time of writing, the MEA is currently out of print and hence difficult to obtain, a comprehensive update of the document is to be issued by the DoT in the near future. (A revised MEA was published as Volume II of the DoT's *Design Manual for Roads and Bridges* in June 1993.)

The DoT Departmental Standard again re-states the criteria for establishing whether EA of a highway proposal is necessary. It also gives further advice on other elements of EA, such as what details should be included in the description of the scheme, the inclusion of proposed measures to mitigate adverse environmental effects and the consideration of alternative alignments.

6.2.4 Other advisory documents

In addition to the official Government advice, there has also been since the introduction of the various EA legislation and regulations the publication of other informal advisory material which may also be applied to the EA of linear development. Apart from 'textbook'-type material for students and practitioners, and articles in trade, professional and academic journals, the documents most likely to be of assistance in relation to the assessment of linear developments are:

- manuals produced as advice to developers by the recipients of Environmental Statements, such as local authorities

- manuals produced by organisations whose members have, in the course of their activities, been called upon to prepare environmental statements.

Examples of these include the general guide to EA produced by Kent County Council (1991) and the guide to assessment of public transport projects produced by the Passenger Transport Executive Group (Lee and Lewis, 1992).

6.3 PRINCIPAL ACTIVITIES IN THE ASSESSMENT

6.3.1 Introduction

In this section, some of the principal activities and tasks that are necessary to be undertaken in carrying out an EA are reviewed and discussed. The section does not deal specifically with the prediction and evaluation of environmental impacts, which are discussed more fully in Section 6.4, but concentrates rather on some of the other aspects of the process that can often be overlooked. These are:

- *the scoping exercise:* identifying the potential sources of positive and negative impact likely to be associated with the project in question

- *the project description:* describing the importance of this within the EA

- *baseline data requirements:* what constitutes the receiving environment for the development in the context of the EA

- *consultations:* how these can be used to inform and enhance the EA

- *presentation:* the need to ensure that the documentation produced during the EA is of the appropriate standard

- *after the ES:* those elements of the EA that continue after the Environmental Statement has been produced.

These are described in turn in the following paragraphs.

6.3.2 The scoping exercise

The purpose of the scoping exercise in any EA project is to ensure that the assessment is focused upon the key environmental issues that are relevant to the type of development to be undertaken and the areas in which it will take place. For transport developments, the scoping exercise becomes particularly important because of the potentially very large geographical extent of the area over which environmental effects may occur. In many cases, the time and resources available to undertake the assessment require conciseness and cost effectiveness to be paramount. In such circumstances, homing in on those features of the infrastructure design that are likely to cause impacts to arise and the location and extent of particularly sensitive receptor groups (people, flora, fauna, special land uses) along the route alignment will best meet the requirement to assess all significant effects.

Basically, the scoping exercise requires a two-stage approach. First, all those physical features of the proposed infrastructure, together with activities associated with its construction and use, which may cause environmental impacts need to be identified. This will involve a detailed examination of the proposals and close liaison and ideally integration with the engineering design team, as it is only a thorough understanding of the nature of these which will allow the identification of the scope and magnitude of environmental impacts. Secondly, the key features of the receiving environment will need to be considered in order to identify the likely interactions between this and the activities associated with the construction and operation of the project. This exercise will also be crucial in identifying some of the key potential areas for the development and application of measures to mitigate adverse environmental effects.

As an example, Figure 6.1 shows in a diagrammatic form the identification of characteristics of a light rail scheme that are likely to give rise to environmental impacts. Such generic approaches to the identification of effects, while useful at the scoping stage, must of course be developed into a specific scheme description relating to the particular development which is under consideration in the assessment. This is discussed further below.

6.3.3 The project description

One of the few specific requirements of the EC Directive and UK Regulations on EA is a description of the project. However, the purpose of such a description within the EA process means that its focus must be the specific features of the proposals being assessed that are likely to result in environmental impacts, the precise nature of these and the disturbance and/or pollutants which are likely to arise. This typically must comprise two parts:

- A general description of the area through which the proposed alignment of the project will pass. This will examine land use, townscape or landscape quality, designated protected areas (conservation areas, Sites of Special Scientific Interest etc.) and sensitive receptors in general terms. The detailed analysis of the baseline environmental conditions will comprise part of the specialist studies that comprise the EA (see Section 6.4).

- A detailed description of the scheme alignment, its main features and associated structures, and the dimensions of these. The capacity of the highway or railway, with predicted traffic flows or service intervals. For rail developments, details will also be required of vehicle specifications (dimensions, braking systems etc.), power sources and equipment. The location and specification of ancillary equipment should also be included. Very often, detailed and accurate information of this type may not be available because the detailed design work may not have been finished (or in extreme cases, not even begun) at the time when the EA is being carried out. Where this is the case, deficiencies in the available information and design features that are likely to be changed should be clearly identified. It may be useful to utilise the concept of a 'design freeze', where it is clearly stated that the EA relates to the design of the project as it stood at a specific date.

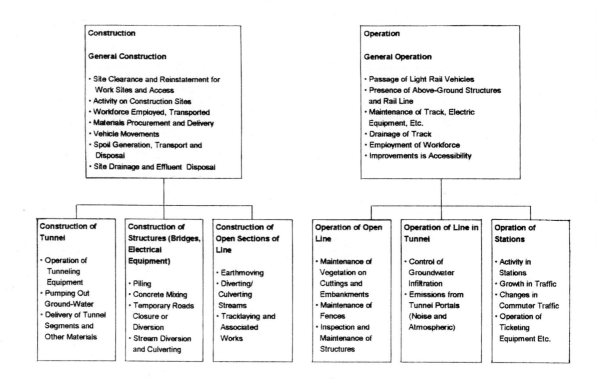

Figure 6.1 *Summary of phases and components of light rail schemes likely to give rise to impacts*

6.3.4 Baseline data requirements

The description of baseline environmental conditions within EA essentially relates to those conditions that would prevail should the proposed development not go ahead. The full description of the baseline therefore requires information not only to be collected in relation to existing conditions, but account must also be taken of other known changes likely to affect these between the present time and the time when the proposed development comes into use. Examples of such changes would be other infrastructure projects already programmed, development proposals (for which planning permission has already been granted, or which are included in development plans), and growth in traffic flows over time. This baseline will have different time horizons when considering impacts associated with construction as opposed to those arising from operation. This will also vary in relation to the consideration of different categories of impact.

Consequently, the collection of baseline environmental data typically is a two-stage process:

- The first stage comprises the collection of data pertaining to existing environmental conditions for the full range of parameters likely to be affected by the proposed development. This might include ambient noise and air quality data, the location and quality of surface and groundwater resources, existing traffic flows and travel patterns, the location and likely extent of ground contamination, data on landscape and visual resources, archaeological features and resources and ecological resources.

- The second stage is to combine this information with an analysis of how environmental conditions will be likely to change over the period of construction and operation of the proposal being assessed. This is done on the basis of information on published plans and committed proposals, together with forecasts of growth in traffic and other demographic and economic indicators.

In practice, most environmental assessment work is undertaken under fairly specific and limited budgetary and time constraints. Therefore, much of the piecing together of the baseline situation will be based on information from published or other readily available sources. For example, information on air and water quality, and on traffic flows and patterns is likely to be available from the routine monitoring work undertaken by local authorities or the National Rivers Authority. For other parameters, however, there are unlikely to be sufficient data easily available in such a manner. Data on ambient noise levels or the qualitative assessment of existing landscape resources are two examples where specific surveys relating to the assessment in hand are almost certainly likely to be necessary.

6.3.5 Consultations

As well as being an important element in the EA process as a whole, and a regulatory requirement, consultations with statutory bodies and other interested parties often form an invaluable part of the data-gathering exercise in establishing baseline environmental conditions. Indeed, the regulations governing EA place a duty upon the statutory bodies who hold such information to provide it to developers undertaking EA at a reasonable cost.

Establishing good contacts with the bodies (statutory or otherwise) has a number of benefits that will help to ensure the quality of the assessment being undertaken. As well as the straightforward task of data collection, it allows access to expertise and knowledge of local conditions which otherwise would be impossible, given the limited time frames within which EA is often undertaken. Setting up such channels of communication to help to obtain baseline data also provides a forum for the discussion (often in relatively informal situations) of specific local issues and concerns relating to the potential effects of the proposed development. This again would usually not be possible within the scope of most EA work, and also helps to make sure the resulting ES covers the full range of pertinent issues.

In relation to major infrastructure projects, consultations associated with the EA may themselves be a major undertaking, requiring commitment and resources from not just those undertaking the assessment, but also from the scheme promoters, statutory consultees and the public. To be effective, the consultations must include a willingness to seek advice from interested parties, to pay attention to what they say and to act upon it. If undertaken properly, the result will be a better-quality assessment and a smoother implementation of the proposals. If mishandled, or entered into half-heartedly, the consultation exercise will only serve to increase suspicion and acrimony from all parties, and could lead to substantial delays in achieving the project objectives.

6.3.6 Presentation

Apart from the fairly basic advice given by the 'Blue Book', there is little guidance available on how the ES resulting from an EA should be presented. As a result, many ES's that have been produced are often lacking in various ways.

The key to producing a high-quality ES lies in keeping in mind the purposes that the document is intending to serve and the audiences that it is seeking to address. To produce and report upon good technical work is not sufficient. Carrying out a good EA is fundamental to the production of a good ES, but it does not guarantee it.

As a minimum, a systematic approach to the assessment must be adopted and reflected in the structure of the ES. Ideally, this should follow the following approach:

- Describe the characteristics of the project.
- Identify the baseline environmental conditions.
- Identify potential sources of impact.
- Set out the criteria for the evaluation of impact significance.
- Predict the extent and magnitude of impacts.
- Evaluate the significance of impacts by reference to criteria.
- Recommend measures for the mitigation of impacts.

It is also important to ensure that the ES complies with legislative requirements, such as the inclusion of a non-technical summary. The production of the ES must therefore include good quality-control procedures. The foremost of these currently available include the EA Review Package developed at the University of Manchester (Lee and Colley, 1990), and guidelines published by the Institute of Environmental Assessment which provide an effective and practical means of helping to ensure that the ES produced is of the highest quality possible.

6.3.7 After the ES

An important point, which is often forgotten, is that the production of the ES does not mark the end of the EA process. There are two key areas where the continued involvement of those producing the ES may be required.

- *Public inquiries and hearings.* Many developments for which EA is undertaken will also need to be the subject of a public inquiry or parliamentary hearing, and this is especially true for linear developments. The material included in the ES may become a topic of debate at such hearings if objections to the proposals are raised on environmental grounds. Also, in such cases, those involved in the EA may be called upon to give expert evidence. It is consequently vitally important that the technical basis of the EA is of the required standard so that the integrity of the work undertaken and the professional competence of those undertaking it cannot be called into question.

- *Monitoring.* The ES is essentially based on predictions, which may or may not reflect what occurs during construction and operation of the proposed development. If the mitigation measures recommended in the ES are to be shown to be effective, the extent and magnitude of environmental impacts that occur during the construction and operational periods should be monitored where appropriate.

6.4 IMPACT IDENTIFICATION, EVALUATION AND MITIGATION

6.4.1 Introduction

In this section, the nature and extent of the likely environmental impacts arising from linear developments are discussed under a number of specific headings. For each of these, the possible sources of impacts are identified both during the construction and operational phase, together with comments on how their magnitude may be predicted. Also, the ways in which the significance of these identified impacts may be established are considered. Then, some of the typical means by which the effects of such impacts may be reduced to acceptable levels are given.

In approaching the general discussion of the environmental effects of linear developments in this way, it must be borne in mind that these do not occur in isolation, and must be considered in relation to who and what will be affected by them. Impacts relating to those issues specified by the EA regulations may be categorised in the following manner. The EA should consider the impacts of linear developments on:

- occupiers and users of facilities in or near the development corridor

- travellers using various modes (particularly in relation to transport infrastructure developments)

- environmental resources (air, water, flora, fauna, landscape resources etc.).

These impacts are discussed below, categorised in the manner set out in Table 6.2.

Table 6.2 Categorisation of impacts arising from linear developments

Main category	Impact	Section no.
Physical and chemical effects	Vibration	6.4.2
	Air quality	6.4.3
	Water quality/hydrology	6.4.4
Natural habitats	Ecology	6.4.5
Human characteristics	Noise	6.4.6
	Severance	6.4.7
	Visual impacts	6.4.8
	Archaeological/historic buildings	6.4.9
	Socio-economic effects	6.4.10

6.4.2 Physical and chemical effects: vibration

The key issues arising from the consideration of vibration effects from linear developments are set out in Table 6.3.

Table 6.3 Examples of vibration impacts from linear developments

Sources of impact	Prediction methods	Evaluation criteria	Mitigation options
Construction			
Piling/tunnelling (inc. blasting)	Computer models	BS 6472	Construction methods
			Changes to alignment
Operational			
Passage of vehicles/trains	Computer models	BS 6472	Design features
			Changes to alignment
			Maintenance

Vibration may occur during the construction of linear developments from the use of tunnel boring or piling operations, particularly in relation to underground works. The use of explosives in excavation work may also be another source. Here again, prediction is difficult given that the exact nature of the plant to be used is often uncertain at the stage when the EA is undertaken. Another important factor is the ground conditions in the location where the construction works are undertaken, as different types of geological strata will propagate vibration at different rates.

Wave propagation in the ground takes several forms. Some waves spread through the ground in a manner analogous to sound waves in air (although both compressional and shear waves may occur). Other waves travel on the surface, more like ripples on the surface of a pool of water. Underground geological features may also cause the reflection and distortion of vibration waves. Another phenomenon that may occur is re-radiated noise. This occurs where vibrations affect structures such as basement retaining walls in buildings, causing them to act as diaphragms. The result may be experienced as a low-frequency rumbling noise.

Vibration from construction works may affect people, wildlife, sensitive equipment and structures. The effects of vibration on people are addressed in BS 6472 (1984), which suggests satisfactory magnitudes of vibration that may be experienced in residential buildings both by day and night. Prior to this, a simpler expression of a peak particle velocity of 12 mm per second was generally accepted as a significance criterion for evaluating vibration effects on humans. BS 6472 also gives criteria for evaluating effects on sensitive equipment, such as that found in operating theatres, although for some proprietary equipment (e.g. computers) manufacturers may specify a criterion themselves. In relation to structures, an accepted limit below which damage to vibration should not occur is given in DIN 4150 as 50 mm per second for impulsive shock waves. However, it should be noted that the limit of perception by humans is much lower than this.

Measures for the mitigation of noise effects, such as the use of bored rather than percussive piling, and the careful consideration of alignments and construction site locations, are discussed further in Section 6.4.6. Apart from these, there is relatively little scope for measures to reduce vibration impacts. However, the likelihood of significant effects occurring in the first place is relatively remote.

Vibration may also arise from the use of linear developments, particularly from the passage of vehicles on highways and railways. This typically takes the form of surface wave propagation set up by the interaction of the wheels of vehicles with the road or rail surface, or by radiating waves where the road or rail is in a cutting or tunnel. While this is unlikely to be a significant effect, there is the potential for such vibration leading to impacts of re-radiated noise.

Criteria for the evaluation of operational vibration effects are broadly similar to those recommended in relation to construction works above.

The mitigation of operational vibration effects chiefly relates to the nature of the wheel surface interactions referred to above. There is some scope for modifications to vehicle design that may assist here. However, although measures such as increasing the number of axles on heavy vehicles are possible, these are obviously beyond the scope of project-specific mitigation recommendations. In respect of railways, the incorporation of vibration-reduction measures is more practicable. Such features may include the use of resilient track mountings to help to absorb vibration. In both cases, maintenance of road or rail surface is perhaps the most important mitigation measure.

6.4.3 Physical and chemical effects: air quality

The air quality issues considered in the section as arising from linear developments are given in Table 6.4.

Sources of air quality impacts that may arise during the construction of linear developments essentially fall into three categories, as follows:

- *Dust emissions:* Dust is generated during construction works chiefly from on-site surface works, including site clearance and preparation, removal of existing surfaces and earth-moving operations. It may also arise from the transporting, use and storage of construction materials (particularly sand, cement and materials for the production of concrete).

- *Contaminated dusts and gases:* Disruption of existing contaminated sites can cause the release of potentially hazardous contaminated dusts and/or gases. Some contaminants may also give off odours when disturbed.

- *Plant emissions:* Emissions from the operation of petrol and diesel-driven construction equipment and vehicles will include nitrogen oxides, hydrocarbons, carbon monoxide and particulate matter.

Table 6.4 Examples of the impact of linear development schemes on air quality

Sources of impact	Prediction methods	Evaluation criteria	Mitigation options
Construction			
Dust emissions	Qualitative evaluation	UK particulates standard	Site management
	Computer models		
Gaseous emissions		EC Standards	
		WHO Guidelines	
Operational			
Dust emissions	Qualitative evaluation	MEA Criterion	Changes to alignment
		UK particulates standards	
	Computer modelling		
Gaseous emissions		EC standards	
		WHO Guidelines	

In terms of evaluating the significance of air quality impacts, it should be noted that for the majority of situations, 70% of dust emissions typically settle within 200 m of their source, and only 10% remain in the air at 400 m away.

No specific guidelines or standards for the levels at which dust deposition becomes a significant problem exist in either UK or EC legislation.

Various US states have established standards varying between 5 and 15 g/m^2 per month, the lower of which is broadly equivalent to 180 mg/m^3 per day. Alternately, the UK national air quality standards for suspended particulates (i.e. smoke) are:

- 80 $\mu g/m^3$ annual median of daily means
- 40–60 $\mu g/m^3$ arithmetic mean of daily means
- 250 $\mu g/m^3$ at the 98th percentile of daily means
- 150 $\mu g/m^3$ arithmetic mean of daily means if measured by a gravimetric method.

However, in practical terms, monitoring such effects in detail is difficult, and although such standards exist generally, their use as evaluation criteria in specific instances is limited.

For contaminated dusts and odours, while many of these compounds have annoyance and human health effects (sometimes extremely serious), these vary according to the degree of exposure in terms of both time and concentration. Very often, there is no safe level of exposure, and again the monitoring of concentrations arising from specific sources is difficult.

In terms of the consideration of such issues in EA, the most practical course of action is a qualitative evaluation of likely incidence of air pollutants and their effects based on expert opinion. Gaseous emissions and pollutant concentrations, however, may be modelled.

For the mitigation of air quality impacts during the construction, the best general approach is to aim to limit the emissions of dust and other materials through good site practice and housekeeping. Such measures might include:

- surface sealing of haul roads

- enclosure of all materials stockpiles, and the use of water sprays to dampen these if required

- design controls for construction equipment and vehicles (e.g. hoods, minimum drop heights for conveyors etc.)

- dust extractor (e.g. from concrete and asphalt plants)

- wheel-washing facilities for road vehicles

- roadsweeping

- sheeting of lorries when transporting materials or spoil

- regular spraying of construction sites and loading/offloading areas for materials or spoil

- minimising double-handling of materials or spoil.

In relation to contaminated dust and odours, any areas of possible contaminated land will need careful investigation prior to the start of construction works. Care must be taken during excavation or disturbance of such material, and friable material (such as asbestos) must only be excavated, transported and disposed of by an appropriately licensed contractor. Other contaminated materials will require, at minimum, disposal to a suitably licensed landfill site or incinerator.

During the operation of linear developments, particularly for transport, air quality impacts will arise from:

- emissions of exhaust gases from the operation of road vehicles or trains

- dust from frictional sources, such as interaction of vehicle wheels with road or rail surfaces, braking systems, or overhead power lines

- possible gas migration from adjacent contaminated land.

Air pollution from road vehicle exhaust emissions may be assessed by methods recommended in the MEA and by the Transport and Road Research Laboratory (Waterfield and Hickman, 1982). This estimates peak carbon monoxide (CO) concentrations as an indicator of overall pollution caused by substances such as hydrocarbons, nitrogen oxides, smoke and lead. CO concentrations are predicted from forecast traffic flows, corrected for speed and the distance of receptors from the road. Computer software based on this methodology is available commercially. Should this graphical screening method indicate that a significant air quality problem may arise, it may be necessary to undertake more detailed analysis, utilising one of the commercially available software packages. Here, the detailed prediction of vehicle exhaust emissions, disposal and resulting concentrations may be specifically modelled for a range of pollutants.

For railways, the pollution from operational emissions depends on the power source used. Diesel-engined locomotives will emit carbon monoxide, nitrogen oxides, hydrocarbons and particulates. Electrically powered trains and light rail vehicles do not give off any exhaust emissions in this way, although small amounts of ozone are generated from arcing between overhead wire or 'third-rail' power sources and the train's power take-off mechanism. However, ozone is rapidly destroyed in the atmosphere, particularly in urban areas.

A broader issue in relation to the air quality effects of electrically powered rail systems is the point of generation (i.e. the power station). This is obviously variable, depending on the fuel used to generate the electricity, such as oil, coal, gas, nuclear energy, biomass, incinerated refuse or renewable sources (hydro, wind or wave power). However, this is generally considered to be beyond the scope of a project EA (see Section 1.4 and Chapter 7).

No specific techniques for modelling the air quality effects of railway operation have been developed, although there are a number of sources of per/km emission factors which may be applied. Emissions modelling, using dispersion models, for generating sources are in widespread use, although they may be considered to be beyond the scope of an EA pertaining to a particular project.

The evaluation of air quality effects from highway schemes may be related to the graphical screening method described above. Within this technique, an air pollution problem is defined as occurring where the average peak hour carbon monoxide concentration is expected to exceed 4 ppm by volume (equivalent to the 8-hour concentration probably being in excess of 9 ppm once a year). If this limit is exceeded, then more detailed analysis is required.

Guidelines and standards set by the World Health Organisation or the EC may be used as criteria for evaluation of the results of any dispersion modelling undertaken. These are set out in Table 6.5.

These may also be used as criteria for the evaluation of air quality impacts identified as occurring from railway operations or power generation.

Table 6.5 Air quality standards

Pollutant	Standard	Averaging period	Comments
Carbon monoxide (CO)	9 pm* 35 ppm*	8 hrs 1 hr	Not to be exceed more than once a year
	25 ppm+	½ – 1 hr	Chosen to prevent carboxyhaemoglobin levels exceeding 2.5 – 3% in non-smoking populations
Nitrogen dioxide (NO_2)	0.17 ppm+	1 hr	Not to be exceeded more than once a month
Lead (Pb)	2 $\mu g/m^3$**	1 year	Not to be exceeded where people are continuously exposed for long periods

* United States Federal Air Quality Standard ** EC Directive, incorporated into UK law
\+ Recommended by the World Health Organisation

6.4.4 Physical and chemical effects: water quality

Table 6.6 sets out the issues discussed in relation to water quality arising from linear developments.

Table 6.6 Examples of the impact of linear development schemes on air quailty

Sources of impact	Prediction methods	Evaluation criteria	Mitigation options
Construction			
Groundwater drawdown or surcharge Contamination Obstruction to ground or surface water flows	Qualitative assessment	EC Directives Control of Pollution Act NRA objectives MAFF recommendations	Choice of construction materials and methods Site management
Operation			
Contamination from runoff	Qualitative assessment	EC Directives Control of Pollution NRA objectives MAFF recommendations	Drainage management

Water quality can be affected during the construction of linear developments in two main ways.

- Groundwater drawdown may occur due to disturbance by construction activities and the presence of new structures or from reduction of infiltration rates and aquifer recharge at spoil disposal sites.

- Groundwater and surface water contamination may occur from effluent seepage, or by the leakage or spillage of other contaminants.

Groundwater drawdown is a particular problem which may be encountered if the project involves subsurface works that extend below the water table. For example, water may be released from the surrounding geological strata into excavations for tunnels, cuttings or pipeline routes, and will require disposal. Such dewatering may lead to drawdown, whereby groundwater levels are reduced in the vicinity of the works. This in turn may cause a reduction in groundwater availability at abstraction wells within the area in which drawdown occurs or may lead to a modification of the soil—water regime. Also, obstruction to groundwater flow may be caused by construction of barriers across an aquifer.

Contamination of groundwater and surface waters may occur from groutings (used in structures to help prevent seepage), oils and lubricants from machinery or other stored chemicals used during construction. This may cause significant problems where construction works are carried out in or near sensitive Aquifer Protection Zones (APZs). Similarly, where works are close to or include bridging over surface watercourses, oil and lubricant contamination may occur, as well as pollution by suspended solids such as cement-production materials. As well as lowering water quality generally, this may alter the biological oxygen demand (BOD) and pH values of watercourses, which may have effects on aquatic ecology. Increased solid loads may also cause additional siltation to occur, with consequent turbidity problems.

Predicting the magnitude of such effects is often difficult given the frequent uncertainty of details concerning construction activities and their precise location at the time the EA is undertaken. Such work therefore commonly has to be undertaken in a qualitative manner, based on expert opinion.

The effects on groundwater may be evaluated on the basis of the following factors:

- the extent and nature of use of groundwaters for potable and other uses in the vicinity of the proposed alignment

- EC Directives on the quality of drinking water (75/440/EEC) and groundwaters (80/68/EEC)

- Aquifer Protection Policies and zones specified by the relevant regional office of the National Rivers Authority as part of the measures implemented by the Control of Pollution Act.

For surface waters, similar sets of criteria may be drawn up taking account of the following:

- the location of downstream sensitive water uses and abstractions

- river quality objectives set by the National Rivers Authority, generally expressed as a National Water Council Quality Classification

- EC Directives on the quality of freshwaters to support fishlife (78/659/EEC) and the quality of bathing water (76/160/EEC), in addition to those identified above in relation to groundwater

- Ministry of Agriculture, Fisheries and Food recommendations for the quality of irrigation water.

Mitigation measures that may be adopted for the avoidance of groundwater drawdown along construction works are generally based on minimising the amount of water entering the excavation workings. This might include:

- the use of grouts of various kinds (although these themselves may be polluting)
- diaphragm walls
- freezing using brine or nitrogen
- well pointing
- pumping
- use of compressed air
- alterations to alignment.

The integrity of finished structures in terms of permeability is essential in avoiding unacceptable levels of seepage and drawdown.

In relation to contamination of groundwater, methods for avoiding this are typically based on two approaches:

- *Groundwater inflow control:* Apart from the minimisation of this as described above, where groundwater inflow does occur, it must be collected into temporary sumps and pumped for disposal via a settlement system before disposal. In this way, seepage of contaminated water back into the groundwater system may be avoided.

- *Pollution control:* 'Good housekeeping' measures should be adopted to minimise contamination, such as the siting of storage tanks for oils and other fluids on impervious bases with bund walls (with valves and coupling within this bund), placing pumps in drip trays, adequate sanitary engineering of employees' facilities and careful plant maintenance to minimise spillages and blow-outs. Where grouts are used, these should be low-toxicity compounds (e.g. containing sodium silicate) in preference to those containing heavy metals or synthetic organics.

In relation to surface water contamination, the above points may be equally applied, with additional measures, such as vehicle-washing facilities and silt traps to help prevent unacceptable build-up of soil and/or oil, spoil and other pollutants entering watercourses by means of runoff.

When linear developments become operational, the main concern relating to water quality is the potential increased contamination possible from increased runoff. Increased areas of embankments and hardstanding also may significantly alter drainage regimes even without increases in pollutants, and may in extreme cases cause problems of flooding if not properly controlled. Consequently, balancing ponds have become a key feature of the drainage regime design of new roads.

The runoff from surface linear developments, particularly for transport infrastructure projects, may contain the following contaminants.

- oils and lubricants caused by leakage from vehicles

- solids such as metals from frictional sources (power source and track wear), braking systems, and the break-up of surfaces from wear over time

- metal particles from corrosion of vehicles or infrastructure

- organic substances such as antifreeze from vehicles, road deiceants (salt) or herbicides used to control track or roadside vegetation

- suspended solids arising from waste materials and refuse deposited on roads or track.

Estimating the incidence and quantities of such contaminants is difficult in specific instances, although comparison may be made with various published studies of the observed magnitude of pollution from such sources from various transport infrastructure developments (ERL, 1991a).

The options for the mitigation of such operational water quality will tend to focus on the details of track or road drainage system design: for example, the incorporation of appropriate pollutant traps. Other options include considering the timing of construction works over or near watercourses to avoid fishing or breeding seasons, and pollution control measures such as booms.

6.4.5 Natural habitats: effects on ecology

The issues discussed in relation to the ecological effects of linear developments are given in Table 6.7.

Table 6.7 Examples of the impact of linear development schemes on ecology

Sources of impact	Prediction methods	Evaluation criteria	Options for mitigation
Construction			
Occupation of land Disturbance Contamination	Qualitative assessment	Presence of statutory designations Rarity Wildlife and Countryside Act	Changes to alignment Fencing off areas Other methods for specific effects (noise etc.)
Operation			
Physical damage Disturbance Contamination	Qualitative assessment	Presence of statutory designations Rarity Wildlife and Countryside Act	Landscape works

During the construction of linear developments, there may be potential for ecological impacts arising from:

- occupation of land for construction works, haul roads etc.

- disturbance to plant and animal communities from dust, release of effluents, noise, vehicle movements and the presence of personnel.

Box and Forbes (1992) identify four main categories of ecological impact from linear development, which are:

- *loss of natural features:* direct loss of habitats, geological exposure or geophysical feature

- *hydrology:* changes in groundwater levels and pollution in runoff

- *other impacts on wildlife:* reduction in breeding success, barrier effects etc.

- *road-led development:* increased urbanisation compounding effects on ecology, additional demand for construction materials etc.

The evaluation of these effects will need to be undertaken by suitably qualified experts in plant and animal ecology, taking account of the likely magnitude of effects on air and water quality. The analysis will also need to consider the spatial extent, intensity and duration of such impacts, together with the extent and quality of affected habitats and the importance of affected species. Such assessments should include close consultation with national and local nature conservation organisations, who will provide invaluable advice on specific local conditions, and in particular seasonal variation that may occur.

Habitats may be assessed according to a range of criteria, such as the extent of naturalness, rarity and diversity. These and others are described in an extensive literature (Ratcliffe, 1977; Usher, 1986).

Existing statutory and non-statutory designations for nature conservation importance and amenity value of sites should also be taken into account.

Species may be assessed on the basis of rarity and the extent to which they are under threat. These and other criteria may be found in publications of the nature conservation organisations (Ratcliff, 1977; RSNC, NCC, 1989). Other factors that may be considered will include the importance of species to wider communities, protection afforded under the Wildlife and Countryside Act 1981 and non-statutory designations for scarcity.

In an urban context, criteria may be applied that relate to the social amenity value of sites or ecological importance owing to the fragmentation of urban wildlife habitats. These include:

- existing ecology and amenity
- local ecological deficiency
- location of wildlife corridors
- potential for ecological and amenity use.

Mitigation measures to reduce the impact of construction works on ecology might include the fencing-off of green space not required prior to commencement. Specific measures to help prevent noise, vibration, air and water pollution from the works are described in Sections 6.4.2 – 6.4.4 and 6.4.6.

During the operation of linear developments, particularly transport infrastructure, the main effects on ecology will again result chiefly from disturbance by noise and effects on habitats arising from air and water pollutant emissions. Methods of evaluating and mitigating these are discussed in Sections 6.4.3, 6.4.4 and 6.4.6. Other effects may include disturbance through air turbulence caused by passing vehicles, physical danger to animals from road or rail traffic and severance.

Box and Forbes (1992) also identify the following categories of mitigation measures for ecological impacts

- changes to alignment
- bridges and tunnels, rather than cuttings and embankments
- tunnels under roads (for badgers, toads etc.)
- drainage
- landscape works
- natural regeneration
- timing of construction programme
- spoil disposal management
- careful consideration of roadstone supply sources
- habitat creation.

6.4.6 Human characteristics: noise

Table 6.8 summarises the issues discussed in relation to noise discussed in this section.

Table 6.8 Examples of the impact of linear development schemes on noise

Sources of impact	Prediction methods	Evaluation criteria	Mitigation options
Construction			
Operation of plant and machinery	Computer modelling	BS 5228	Construction methods
Piling/tunnelling Blasting		DoE Advisory Leaflet 10/73	Changes to alignment
Operation of vehicles		WHO	Screening
			Insulation
Operational			
Passage of vehicles/trains	Computer modelling	Modelling Insulation Regulations 1975	Reduction
		'Mitchell' Report	Changes to alignment
			Maintenance
			Screening
			Insulation

Any construction project necessarily involving the use of heavy plant and machinery will cause nuisance from noise to a greater or lesser degree. In relation to EA, the predictions of such effects often have to be made before the construction contractor has had the opportunity to specify in detail their proposed methods of working, or even more usually, before a contractor has been appointed. It is therefore often the case that it is impossible to state precisely what equipment will be used on site.

However, many linear developments have obviously already been constructed, and it is possible, based on experience from these, to suggest the types of equipment that would be present on site at one time or another. These might include:

- earth-moving plant such as excavators, dozers, dump trucks etc.

- surface-breaking equipment, for example pneumatic drills, hydraulic hammers, rock breaking equipment, etc.

- use of explosives for rock blasting

- piling plant or tunnelling equipment

- lifting equipment including cranes and hoists

- concrete and asphalt manufacturing plant such as lorries, mixers, pumps etc.

- miscellaneous equipment, including compressors, hand tools, generators, mixers, lorries etc.

BS 5228 (1984) gives an indication of noise levels from various items of plant taken from measurements at various sites. From this information, it is possible to estimate noise levels emitted by such equipment.

In terms of predicting the specific magnitude of noise impacts from construction works, computer models are available that enable these typical levels to be corrected for specific site conditions and thus allow prediction of facade noise levels likely to be experienced by, for example, nearby residents during the construction period.

One of the major difficulties encountered in predicting noise impacts from the construction of linear developments, however, is that construction activity is often temporary and mobile rather than at fixed and relatively permanent locations. The reasons for such difficulties include the following.

- The works are of a nature where rigorous noise control measures may be restrictive and result in an unreasonable prolonging of the site programme.

- Work sites are not fixed, and will change according to the demands of the construction programme.

- Much of the works are conducted out of doors without the benefit of fixed plant houses.

- A large amount of mobile plant is used.

- Maintenance work needs to be carried out.

In relation to the development of criteria to assess the significance of construction noise impacts, a number of sources of advice are available. BS 5228 lists a number of factors affecting considerations of the acceptability of site noise:

- site location
- existing ambient noise levels
- duration of site operations
- hours of work
- attitude of the site operator
- noise characteristics of the work.

Also, Department of the Environment Advisory Leaflet No. 72 (although now no longer in print) gives advice on the general limits of acceptability of construction site noise during the daytime, stating that this should not exceed 75 dB(A) outside the nearest occupied room (i.e. workplace or dwelling) in rural, suburban or urban areas away from main road traffic or industrial noise.

Both BS 5228 and the DoE Advisory Leaflet suggest that, in the evening, a noise level 10 dB(A) below the acceptable daytime level is appropriate. At night, BS 5228 implies that acceptable noise levels would be those that do not disturb sleep at the nearest dwelling. In addition to this, DoE Circular 10/73 (DoE, 1973) suggests that a maximum acceptable level is 45 dB(A), while a 'good standard' is 35 dB(A) (both after any corrections for tonal or impulsive noise). Also of note is the 35 dB(A) L_{Aeq} level recommended by the World Health Organisation as a level at which the restorative process of sleep is preserved (WHO, 1980).

Table 6.9 summarises these suggested criteria for construction noise evaluation.

Table 6.9 Criteria for evaluating the significance of noise during construction (external levels)

Period	Location	Criterion	Purpose
Day (0700–1900)	Dwellings/offices (facade)	75 dB(A), 12 hour L_{Aeq}	To maintain speech communication
Evening (1900–2200)	Dwellings (facade)	65 dB(A) 3 hour L_{Aeq}	To maintain speech communication
Night (2200–0700)	Dwellings (facade)	40–45 dB(A) 1 hour L_{Aeq}	To avoid sleep disturbance
Day (0900–1600)	Sensitive locations (e.g. schools)	60 dB(A) 1 hour L_{Aeq}	To maintain speech communication

In terms of minimising the effects of noise during construction, it is possible to use noise control measures that do not unduly inhibit work patterns and to employ working methods that result in the minimum noise impact compatible with normal practice. This is generally referred to as the 'best practicable means' approach to reducing noise emissions.

Examples of the sort of measures which may be employed to help mitigate the effects of construction noise by the adoption of a best practicable means policy may include:

- screening or enclosure with acoustic panels of fixed plant, such as pumps or compressors

- scheduling unavoidably noisy operations, such as initial site clearance, to avoid disturbance to residents or users of facilities nearby

- blasting only during specified periods

- use of electronically operated plant in preference to diesel plant where possible

- provision and maintenance of effective silencing to motorised plant

- use of inherently quieter techniques (e.g. bored pilings) where practicable.

In addition, the more flexible and mobile nature of construction practices does often give some scope for including environmental considerations in the selection of primary work sites, for example ensuring that these are situated away from residential areas.

In the operational phase, there is a probability that noise may not be a problem at all, where, for example, a pipeline or urban railway is located underground. Where operation takes place on the surface, however, noise is the environmental impact most often perceived as a nuisance. Prediction methods for road traffic noise are well established and widely used (DoT, 1987), and are also now readily available in the form of computer software. In relation to railway noise, while similar standardised methods are not yet available, there nevertheless has been a good deal of research undertaken and reported in this field, both in respect of conventional and light railway systems.

For both highway and railway developments, however, predictions of noise impacts will rely on the availability of forecasts of traffic flows or operational schedules. Much uncertainty will often be attached to these at the time the EA is undertaken.

In terms of evaluating the significance of transport noise effects, there are readily available criteria that may be adopted. For new highways, the provisions of the Noise Insulation Regulations 1975 apply. These provide criteria of acceptability above which noise attenuation measures must be provided to dwellings affected by traffic noise where:

- dwellings experience a $L_{10,18hour}$ noise level (i.e. between 0600 and 2400) of greater than 68 dB(A)

- at least a 1 dB(A) increase in the $L_{10,18hour}$ noise level is predicted

- at least 1 dB(A) of any such increase is directly attributable to traffic using the new road.

No such legislative standards for evaluating the effects of railway noise currently exist in the UK, although this issue has been addressed by a Committee set up by the Department of Transport. The Committee recommended (DoT, 1991) that noise attenuation measures should be provided to dwellings where the facade noise level from trains operating on a new line would exceed 66 dB(A) L_{Aeq} over a 24-hour period or 61 dB(A) L_{Aeq} between 2300 and 0700, and there has been an increase of railway noise of at least 1 dB(A). In responding to the Committee's recommendations, the DoT have accepted a slightly modified form of these criteria, suggesting that insulation be provided at levels of 68 dB(A) L_{Aeq} for the daytime period 0600 to midnight and 63 dB(A) for the night-time period midnight to 0600 (DoT, 1992).

The DoT are expected to issue formal recommendations on these criteria before the end of 1993. (DoT published Draft Railway Noise Regulations, including a proposed prediction method, in October 1993.)

In respect of the mitigation of operational noise from highways and railways, as may be seen from the above, most emphasis has been placed on the provision of noise attenuation measures to dwellings. This typically comprises fitting double windows (with at least a 10 cm gap between windows) together with ventilation equipment and in some cases venetian blinds. Other options for the mitigation of operational noise impacts may include:

- the consideration of the location of sensitive receptors at the route planning stage, so that the alignment may be placed away from these, thus minimising the potential for noise impacts

- the consideration of changes to the vertical alignment, such as placing highways or railways in tunnels or cuttings, or avoiding the use of elevated sections, specifically to avoid noise impacts

- the use of acoustic screens, noise barriers or earth bunds to reduce noise impacts. This may include using non-sensitive buildings (e.g. warehouses) in a barrier role, or could involve new developments decking over roads or railways

- the incorporation of specific design features, such as the use of porous asphalt surfaces for roads or avoiding tight curves and incorporating the use of embedded track for railways. There is also scope for additional design to trains (e.g. resilient bogie designs) to reduce noise at source

- ensuring good maintenance of road and rail surfaces to prevent increases in noise levels arising through wear to these, if appropriate.

6.4.7 Human characteristics: severance and other traffic effects

The traffic issues associated with linear developments are discussed in this section and summarised in Table 6.10.

Table 6.10 Examples of traffic effects associated with linear developments

Source of impact	Prediction methods	Evaluation criteria	Mitigation options
Severance	MEA techniques	MEA	Change of alignment Footbridges etc. Financial compensation
Traffic noise	Computer modelling	Noise Insulation Regulations	
Air pollution	Computer modelling	MEA	
Accidents	Qualitative assessment	–	
Increased delays	Computer modelling	–	

Severance is an effect that is often associated with linear development projects, resulting either from the presence of the new infrastructure itself or from new and/or increased vehicle flows. Enforced changes to routes or additional delays can increase journey times for travellers having to cross new or more heavily used routes either on foot or in vehicles, and may even deter some journeys from being made altogether. Severance may also have an effect on land uses by causing alterations to catchment areas of commercial uses (shops etc.) or community facilities (e.g. schools, libraries) and thereby affecting their viability. Such effects will tend to apply equally during both the construction and operational phases of a linear development.

Methods and evaluation criteria for the assessment of severance effects are given in the MEA and also in a more recent TRRL contractors report (Clark et al., 1991).

The mitigation of severance effects may be assisted by three types of measures:

- at the planning stage, through the careful choice of route alignments taking account of land use patterns, catchment areas and ownerships

- at the design stage, by the incorporation of features to help to retain existing journey patterns (such as footbridges and underpasses) or land-use patterns (through the organisation of land-transfer arrangements)

- through the adoption of a compensation scheme to assist financially those affected by severance effects.

Other impacts may also arise from changes in travel patterns due to linear development projects. These include change to the levels of:

- traffic noise
- air pollution
- accidents
- increased delays to other road users.

The prediction of such effects relies on the adequate modelling of the changes in journey patterns likely to occur during both the construction and operational phases. This may be available through a separate traffic impact study undertaken in relation to the specific project under consideration. On the basis of this information, the consequent environmental effects may be predicted utilising the techniques described elsewhere in this chapter.

The mitigation options for minimising such off-site traffic effects are limited, but may include the provision of additional improved road infrastructure through planning agreements and the development of a routeing strategy for construction traffic that seeks to avoid sensitive receptors.

6.4.8 Human characteristics: visual impacts

The key visual impacts discussed in this section in relation to linear developments are set out in Table 6.11.

Table 6.11 Examples of visual effects associated with linear developments

Sources of impact	Prediction methods	Evaluation criteria	Mitigation options
Blocking views	Qualitative assessment	Qualitative assessment	Change to alignment
Opening up views	(MEA, Countryside Commission)		Quality design
Removal of important features	Photomontage		Landscaping
Introduction of new elements	Models		

The assessment of the visual impact arising during construction and use of linear developments is essentially a subjective exercise. However, there are accepted methods available to help to provide a standardised approach to such qualitative assessments, such as those described in the MEA or the PTEG Guide. A vital element in undertaking such assessments is that they must be undertaken by a qualified landscape architect or urban designer.

Typically, the key tasks in the assessment of visual impacts are as follows:

- *Defining visibility:* This is the definition of the potential extent, or the area over which impacts may be experienced, and sets out the geographical limits within which receptor sensitivity and landscape resources are assessed. Visibility is normally determined by distance or physical obstructions such as topography and buildings.

- *Defining receptor sensitivity:* Potential receptors within the zone of visual influence (as defined above) may be classified by their numbers and sensitivity. Sensitivity is influenced by a number of factors such as distance, viewing opportunity and receptor interest.

- *Defining landscape resources:* This is in effect the identification of the baseline visual conditions that may be affected by the proposal.

- *Prediction and evaluation of impacts:* This involves identifying changes such as the obstruction and intrusion into existing views, opening up of new views and the qualitative changes to existing views. Other factors to be considered will include the loss of visually significant features (e.g. mature trees or historic buildings), changes in land use patterns or the quality of urban space and the introduction of features that may conflict with their setting.

The mitigation of visual impacts will be a key element in the overall design process of a linear development, and indeed it is usually the case that comprehensive landscape and/or townscape design proposals are a part of the overall project. Typically, such designs will focus on four main elements in attempting to reduce the visual effects of a proposal:

- consideration of screening etc. during construction
- care and quality in the design of key visual elements in the proposals
- careful consideration of the design of ground-level components
- remedial measures such as screening by dense planting and mounding.

Also through the development of landscape design proposals, positive contributions may be made to the mitigation of ecological impacts (see Section 6.4.5).

6.4.9 Human characteristics: effects on archaeology and historic buildings

The potential sources of impact upon features and sites of historic and archaeological importance associated with the construction of linear development projects are summarised in Table 6.12.

Table 6.12 Examples of impact of linear development schemes on archaeology and historic buildings

Sources of impact	Prediction methods	Evaluation criteria	Mitigation options
Construction			
Land take	Qualitative assessment	DOE PPG 16	Change to alignment
Vibration/settlement			Preservation *in situ*
Groundwater drawdown			Excavation and record
Land-use change			
Remedial measures			
Operational			
Direct damage	Qualitative assessment	DOE PPG 16	Change to alignment
Emissions			Preservation *in situ*
Vibration			Excavation and record
Land-use change			
Amenity loss			

The following detailed points relate to the potential sources of impact on archaeological and historic measures:

- *Land take:* Surviving archaeological deposits near the surface generally, or in deeper strata where tunnels or cuttings are proposed, may be compromised or destroyed by construction works. Similarly, there may be direct effects on historic buildings, upstanding archaeological remains or important townscape features.

- *Vibration and settlement:* Construction vibration from excavations, piling or tunnelling are effects to which some historic buildings may be especially vulnerable. Where substantial surface excavation or tunnelling is proposed, there may also be effects on historic buildings as a result of settlement.

- *Groundwater drawdown:* The lowering of groundwater levels due to excavation or tunnelling works (see Section 6.4.4) may affect the integrity of waterlogged archaeological remains preserved *in situ*.

- *Land use:* Neglect arising from changes in land use and management (e.g. farming practices). Effects on historic landscapes.

- *Damage caused by remedial measures:* Soundproofing of buildings and measures taken to remedy subsidence (such as shoring or underpinning) can cause damage to the historic or architectural integrity of buildings. Also, sound barriers, embankments, bridges etc. may damage or otherwise affect historic sites, features or buildings and their settings.

In the assessment of such impacts, there are no standard scales of comparison against which they may be determined, owing to the great diversity of features and their different vulnerability due to the range of likely impacts. However, a common approach to the prediction of the magnitude and extent of impacts may involvement consideration of:

- the number of features affected

- the importance of these (see below)

- guidelines provided by specialists on the likely effects of vibration etc. either in general or in site-specific terms.

The significance of impacts on historic or archaeological features may be evaluated by reference to non-statutory guidelines produced by the Department of the Environment (1990). This suggests that the importance of archaeological and historic features may be evaluated by the consideration of:

- survival/condition
- period
- parity
- fragility/vulnerability
- diversity
- documentation
- group value
- potential.

This evaluation must be undertaken on the basis of competent professional judgement, and is a particular area where consultation with the regional officers of English Heritage (or its equivalent in Scotland or Wales), local authority archaeological offices and local heritage organisations is vital.

Similarly, in the development of measures to mitigate effects on archaeological and historic features, it is essential that the national and local heritage organisations are involved. The detailed historic framework of the area will need to be established, and perhaps a programme of field trial works drawn up. On the basis of this, together with expert advice, appropriate mitigation measures may be developed. For features of importance, preservation *in situ* will be the preferred course of action. Elsewhere, it may be acceptable to preserve the historic or archaeological information through controlled and fully recorded excavation.

After completion of construction works, the potential for impact on historic or archaeological impacts is reduced. However, adverse effects may still occur, particularly in relation to transport projects, and may include:

- *Direct damage:* Physical damage may occur through accidental incidents (generally a very rare occurrence).

- *Deposition of emissions:* Compounds from vehicle emissions may stain or damage stonework of historic buildings through direct deposition or in solution in precipitation.

- *Vibration:* Vibration from the passage of vehicles may affect the structural integrity of sensitive historic buildings (although the occurrence of this is rare).

- *Dereliction or neglect:* Buildings in close proximity to the proposals may suffer decay through dereliction and neglect due to abandonment or change of use. However, this is inevitably in most cases beyond the control of the promoter.

- *Loss of amenity:* The value of a particular feature or site may be affected by noise or visual intrusion.

The evaluation of these effects and the development of measures to mitigate them will broadly follow that described above in relation to the effects of construction works.

6.4.10 Socio-economic effects

The issues discussed in this section are identified in Table 6.13.

Table 6.13 Examples of socio-economic effects of linear developments

Sources of impact	Methods of prediction	Evaluation criteria	Mitigation options
Construction			
Land take	Quantitative and qualitative assessment	Structure, local or unitary development plan policies	Minimising land take choice of alignment land swaps Maintaining access Financial compensation Developing appropriate policy responses
Severance			
Accessibility			
Employment			
Effects on community facilities			
Operation			
Travel patterns	Qualitative assessment	Structure, local and unitary development plan policies	Developing appropriate policy responses
Accessibility			

During the construction of linear developments a number of socio-economic impacts may arise from both the occupation of land for the construction works and in relation to the construction activities themselves.

- *Land take:* The occupation of land may lead to immediate impact, forcing out existing uses and/or resulting in demolition of buildings. In addition to the permanent land-take required for the road, railway or pipeline installations themselves, land will also be required either for temporary occupation for construction sites, or through easements and/or other agreements for access, particularly where works are underground in bore or cut-and-cover tunnels. Neighbouring land uses may also be subject to modification or change due to the physical proximity of the new development.

- *Severance:* Mention was made in Section 6.4.7 of the possible severance effects on the viability of commercial uses and community facilities. Severance may also have a more direct effect, where individual sites, either currently in use or planned for redevelopment, are bisected by the route of the road, railway or pipeline. This may lead to the remaining sites becoming unviable for their current or intended use. Consequently a differing use may occur, or the land may fall into disuse.

- *Accessibility:* Change to patterns of access, particularly the closure of roads, may have detrimental effects on some land uses and force changes to occur. Similar effects may also occur through increases in traffic flows caused by traffic generated from the construction works.

- *Employment:* The construction works may be a major source of new employment, and in some cases may make a significant contribution to local economies through local multiplier effects. Supply contracts placed with local firms may also provide a boost to local economies. Conversely, some adverse effects on the health and diversity of local economies may also occur.

- *Effects on community facilities:* It is now not uncommon for major construction projects to attract workers from far afield, and temporary accommodation is often provided for these people (e.g. in the form of caravan parks or in local hotel accommodation). The influx of construction workers into an area may have a number of effects, both positive and negative. The financial effects may benefit the local economy as noted above, and there should be some opportunities of employment for local people. However, additional pressures may be placed on local community services, such as housing, schools etc. Where temporary accommodation for construction workers is provided, there may be additional problems of opposition from local residents, especially where tourism is important in the local economy.

The measurement and evaluation of such effects will require subjective analysis by a skilled economist, chartered town planner or chartered surveyor. The consideration of direct effects on existing land uses will require consultation with existing occupiers and an assessment of how the viability of activities may be affected. The potential effects or planned uses will also need to be taken into account. The secondary economic effects may be estimated by examination of the characteristics of the local economy, together with an analysis of how the injection of new money distributed through wages and contracts with local firms may affect these.

The predicted impacts of the construction works on land use, community facilities and the local economy may be evaluated by reference to the structure and local plans (or Unitary Development Plan) produced by the relevant local authorities for the area at county, district or metropolitan borough level. Should the effects of the proposed development substantially prejudice the achievement of the policy goals set out in such plans, then it may be said that a significant adverse impact will occur. However, one should not lose sight of the fact that even where specific policy goals are not affected, the loss of a particular use (e.g. commercial activity with subsequent job losses, or the loss of homes) may be significant in its own right and should be identified as such.

The scope for developing project-specific mitigation measures in relation to socio-economic effects is limited, but may include the following:

- *Minimising land take:* Ensuring that the design of the project occupies as little land as possible.

- *Alignments:* Choosing an alignment for the proposal that takes account of existing land use patterns, so as to minimise disruption and severance of these.

- *Land swaps:* Negotiation with land owners to include arrangements for the transfer of parcels of land severed or otherwise affected by the proposals. This will help to minimise direct impacts of the proposals by assisting the maintenance of the viability of existing land uses.

- *Maintaining access:* The provision of physical links (e.g. bridges or tunnels) to maintain existing vehicular and pedestrian routes. New links may also need to be required to offset severance effects.

- *Financial compensation:* Financial recompense to landowners and homeowners for effects on business arising from both physical and other impacts.

More generally, the proposals should ideally form part of an overall approach to land-use planning. New infrastructure projects such as roads and railways will often have been included in structure and local plans as part of the processes and procedures of gaining the appropriate consents. However, where the promoters of a particular scheme are not also a planning authority, close liaison will need to be maintained between the two parties to develop the appropriate responses. There is also scope for entering into specific agreements with the local authorities under the Planning and Compensation Act 1991 for the provision of some of the measures described above. The local community should also be involved through the consultation programme as described in Section 6.3.5.

When the development becomes operational, particularly for transport infrastructure projects, there may be more fundamental effects on land use and activity patterns. These are essentially two-fold:

- *Changes to travel patterns:* New transport infrastructure will inevitably lead to a rearrangement of travel patterns. Some journeys may shift to the new road or railway, and new feeder trips may also be generated as a result. Some completely new journeys may also be generated. Modal shifts may also occur. Such factors may in turn have indirect effects on land-use patterns and economic activity

- *Changes in accessibility:* The new transport infrastructure will improve accessibility to some sites and lead to an increase in development potential and consequent increased pressures for development. Similarly, access may be worsened to some sites, with longer-term effects on the viability of existing uses.

The measurement and evaluation of such effects, for those during the construction period, must be undertaken by appropriately qualified people in the context of the stated planning and development policies for the area to be affected.

The development of mitigation measures to address the operational effects described above can only effectively be done through the media of structure, local or unitary development plans and the development control process. Here again, where the promoter of the project is not also the planning authority, close liaison must be maintained with them (and the local community through public consultation) to help the development of the appropriate policy responses to help maximise the longer-term economic and social benefits of such proposals.

6.5 REFERENCES

BOX, J.D. and FORBES, J.E. (1992)
Ecological considerations in the environmental assessment of road proposals
Journal of the Institution of Highways and Transportation, Vol. 39, No. 4, pp 16−22

BS 5228: Part 1 (1984)
Noise control on construction and open sites
British Standards Institution

BS 6472 (1984)
Evaluation of human exposure to vibration in buildings
British Standards Institution

CLARK, J.M. *et al.* (1991)
The Appraisal of Community Severance
Transport and Road Research Laboratory, CR 135

DEPARTMENT OF THE ENVIRONMENT (1973)
Planning and Noise
Circular 10/73, HMSO

DEPARTMENT OF THE ENVIRONMENT (1990)
Planning Policy Guidance 16: Archaeology and Planning
HMSO

DEPARTMENT OF TRANSPORT (1987)
Calculation of Road Traffic Noise
HMSO

DEPARTMENT OF TRANSPORT (1992)
The Government's Response to the Mitchell Report

DoT DEPARTMENTAL COMMITTEE ON RAILWAY NOISE (1991)
Railway Noise and the Insulation of Buildings (the Mitchell Report)
Department of Transport, HMSO

KENT COUNTY COUNCIL (1991)
The Kent Environmental Assessment Handbook
Planning Department

LEE, N. and COLLEY, R. (1990)
Reviewing the quality of environmental statements
Occasional Paper No. 24A, EIA Centre, Department of Planning and Landscape, University of Manchester

LEE, N. and LEWIS, M. (1992)
Environmental Assessment Guide for Passenger Transport Schemes
Passenger Transport Executive Group

PERRING, F.M. and FARRELL, L. (1983)
British Red Data Book

RATCLIFFE, D.A. (1977)
A Nature Conservation Review Vol. 1
Cambridge University Press

RSNC, NCC (1989)
Guidance on Selection of Biological Sites of Special Scientific Interest
Royal Society for Nature Conservation and Nature Conservancy Council

USHER, M.B. (1986)
Wildlife Conservation Evaluation
Chapman & Hall

WATERFIELD, V.H. and HICKMAN, A.J. (1982)
Estimating Air Pollution from Road Traffic: A Graphical Screening Method
Transport and Road Research Laboratory SR 752

WORLD HEALTH ORGANISATION (1980)
Environmental Health Criteria 12: Noise

Apart from the specific references given above, the following sources have also been used in preparing this chapter:

COLLEY, R. and FERRARY, C. (1991)
Environmental assessment of highways: the EC Directive and current practice
PTRC Summer Annual Meeting, Proceedings of Seminar L, pp 25–32

ENVIRONMENTAL RESOURCES LTD (1989)
Channel Tunnel rail link environmental assessment: scoping report
British Railways Board

ENVIRONMENTAL RESOURCES LTD (1990)
Central area transmission system onshore gas pipeline: environmental statement
Amoco (UK) Exploration

ENVIRONMENTAL RESOURCES LTD (1990)
Greater Manchester Light Rapid Transit Bill 1989: environmental statement
Greater Manchester PTE

ENVIRONMENTAL RESOURCES LTD (1990)
Greater Manchester Light Rapid Transit Bill 1990
Greater Manchester PTE

ENVIRONMENTAL RESOURCES LTD (1990)
Jubilee Line extension: environmental statement
London Underground Ltd

ENVIRONMENTAL RESOURCES LTD (1990)
Midland Metro Bill 1990: environmental statement
Centro (West Midlands PTE)

ENVIRONMENTAL RESOURCES LTD (1991a)
Freshwater Pollution Impact Research Study III: Transport
Royal Commission on Environmental Pollution

ENVIRONMENTAL RESOURCES LTD (1991b)
Leeds Supertram Bill 1991: Environmental Statement
West Yorkshire PTE/Leeds City Council

ENVIRONMENTAL RESOURCES LTD (1991c)
Midland Metro Bill 1991: environmental statement
Centro (West Midlands PTE)

FERRARY, C. (1990)
Environmental assessment and transport
Journal of the Royal Town Planning Institute, Vol. 76, No. 44

FERRARY, C. (1990)
Environmental assessment: the transport element
Journal of the Institution of Highways and Transportation, Vol. 37, No. 11

FERRARY, C. (1991)
Environmental assessment of light rail projects
Paper to Light Rail Transit 91 Conference, Aston University

CIRIA Research Project 424

Environmental Assessment

A guide to the identification and mitigation of environmental issues in construction schemes

7. Electricity generation

A Robson, Nuclear Electric plc
T.L. Shaw, Shawater Ltd
C.J.L. Taylor, Nuclear Electric plc

Note: The Nuclear Electric contribution is made with the permission of the company; the opinions presented are those of the authors.

7 Electricity generation

7.1 INTRODUCTION

This chapter reviews the procedures used for carrying out environmental assessments (EAs) and preparing Environmental Statements (ESs) for electricity-generating projects.

The chapter concentrates on outlining the principles involved in assessment and mitigation, with illustration of impacts and detail of assessments from fossil, nuclear and renewable systems for which commercial projects are in operation or proposed.

Because much of the development of impact assessment methods has taken place in relation to large combustion systems, these provide a main source of the examples and case studies quoted herein. While the potential level of impact may be different, the principles and procedures are equally applicable to smaller systems.

For technologies not yet available for wide commercial application, the potentially most significant impacts are noted. These include tidal, biomass, solar, advanced fossil and geothermal systems. When these technologies become more commercially viable, the principles and general methods of environmental assessment outlined herein will be equally applicable, although the emphasis, i.e. which are the most significant areas of impact, will differ from system to system.

Methods for the preparation of EAs for thermal power plants are relatively well established. The UK's electricity supply industry found an EA-type approach and its documentation to be an aid to methodical planning even before legislation made them obligatory.

The orderly development of a project benefits from the parallel implementation of the EA process, especially if used to identify means of mitigating perceived adverse effects. It also provides for the early optimisation of choices between possible locations for the generating plant, water intakes and outfalls or for alternative routes for pipelines and transmission lines. The early broad scoping of environmental issues will also help to establish where and when more detailed studies will be necessary.

A comprehensive EA provides a clear picture of existing, baseline environmental circumstances. Integrating this EA with scheme design particulars and operational processes allows identification of the most significant areas of potential impact. The construction phase will be shorter than the operational phase and will involve different sorts of impact, but both phases warrant similar standards of environmental appraisal. Part of the impact of any project will arise from its related transport activities.

The structure of this chapter follows the lines set out in Chapter 1. Section 7.2 outlines the general legal framework that applies to EAs for electricity-generating plant. Legal controls of a more general nature are drawn together in Chapter 2.

The concept of integrated pollution control (IPC), now widely regarded as basic to environmental regulation in the electricity (and other) industries, is introduced in Section 7.3 and is followed up in more detail in subsequent sections in this chapter.

Legislation is given in greater detail under the separate headings of air, water, etc. in Section 7.4 and subsequent sections, but it must be remembered that this legislation and the regulatory system are subject to change. This has been particularly evident in the UK following the Environmental Protection Act 1990, and is expected to continue with the development of integrated pollution control, the introduction of environmental quality standards and objectives, and coastal zone management.

Environmental regulations in the UK are influenced by the activities of the European Community. Since the early 1980s, the EC has been responsible annually for some 20–30 items of environmental legislation of one sort or another, some affecting the electricity industry.

While the present chapter aims to describe the basic legislation from which the various detailed regulations are derived, it must be the responsibility of the individual developer to ensure that at any time the necessary regulations are understood and adhered to.

Following Sections 7.7 (Noise and vibration) and 7.8 (Visual impacts), topics of a more source-specific nature are considered. Ecological issues not otherwise covered are addressed in Section 7.12, and this is followed by brief reference to the several currently smaller contributors to UK energy supplies, such as biomass and hydro, now receiving wider attention.

7.2 KEY LEGISLATION ON EAs FOR ELECTRICITY GENERATION

7.2.1 Legal framework

Industrial developments have to comply with a wide range of planning and environmental regulations. As noted, these regulations are continually evolving. The guidance given here covers the main sources of legislation in the various general areas of interest. It is the responsibility of developers to ensure that their development complies with these requirements. Updates of legislation are available from many sources. Useful ones in the UK include the *CBI Environment Newsletter*, the *Pollution Handbook* published annually by the National Society for Clean Air, the *Marine Pollution Bulletin* and the Environmental Data Service (ENDS). Various consultancies, legal firms, and publishers offer environmental briefing/information services.

Regulations and guidelines relevant to EAs in the UK derive from the EC Council Directive dated 27 June 1985 on the Assessment of the Effects of Certain Public and Private Projects on the Environment (85/337/EEC). The EC has also proposed a Convention on Environmental Impact Assessment in a Trans-boundary Context (COM(92)83 Final, March 1992). The Department of the Environment (DoE) and Welsh Office (WO) have published *Environmental Assessment: A guide to the procedures* (DoE/WO, 1991a). This contains (in its Appendix 8) a list, revised to August 1990, of UK statutory instruments and other official publications on environmental assessment.

An important development in UK environmental protection has been the Environmental Protection Act 1990. This stresses the 'polluter pays' principle, the concept of integrated pollution control (IPC) and the development of emission controls via the philosophy of BATNEEC (Best Available Techniques Not Entailing Excessive Cost). It also emphasises the duty of care for solid wastes. The Act is being implemented by means of more detailed regulations and guidance such as the Code of Practice on Duty of Care in the Management of Controlled Waste (DoE, 1990a).

The applications under Environmental Protection (Applications, Appeals and Registers) Regulations 1991 require supporting detail on the environmental consequences of any discharges.

Authorisations relevant to environmental assessment are covered by the Environmental Protection (Prescribed Processes and Substances) Regulations 1991 (SI 1991/472 and 1992/614), which define processes and substances scheduled for control by HM Inspectorate of Pollution (HMIP), or local authority(ies). Part A processes are bigger than Part B, have potentially greater impact and require IPC. For Part A projects, HMIP is responsible for IPC and Consents for discharges to air, water and land.

The report *Integrated Pollution Control: A practical guide* (DoE/WO 1991b, updated in 1993) and a series of Process Guidance Notes (IPR series) issued to Inspectors by the Chief Inspector, HMIP, deal with assessment of applications for Consent to operate different sorts of plant. The Guidance Notes include *Industry, Fuel and Power – Combustion Processes*, e.g.

IPR 1/1 for large boilers and furnaces 50 MW(th) and over and IPR 1/2 for gas turbines. Various General Guidance Notes cover areas such as *Authorisations* (General Guidance Note 2) and *Applications and Interpretation of Terms Used*.

Part B processes are those where local authorities deal with discharges to air (only), i.e. small boilers, gas turbines, waste oil burners, waste-derived fuel incinerators and other combustion systems. For local authority consents, the DoE/WO issue appropriate Secretary of State's Process Guidance Papers, e.g. for combustion boilers 20−50 MW and for waste-derived fuel up to 3 MW(th).

Demonstration of the ability of a project to comply with these guidance requirements, or the intent that they should, provides a useful part of an EA, as too does demonstration of an acceptable impact relative to environmental quality standards and objectives for the area around the proposed plant.

Applications to construct, extend or operate an electricity-generating station having an installed capacity of more than 50 MW are made under Section 36 of the Electricity Act 1989, to the Secretary of State (now the Secretary of State for Trade and Industry). Applications to install or keep installed an electric line above ground are made under Section 37 of this Act. The Electricity (Applications for Consent) Regulations 1990 outline requirements such as those for publication of notices for consent and introduce an obligation for such notices to be served on the Nature Conservancy Council (since replaced by separate bodies for England, Scotland, Wales and Northern Ireland and hereinafter referred to collectively as 'English Nature and its equivalents') where the land in question is a designated Site of Special Scientific Interest.

The formal views of the local planning authority are obtained through a statutory procedure and are considered by the Secretary of State in arriving at his decision.

Proposed use of land other than for agricultural or defence purposes requires either planning permission from a local authority or deemed planning permission. If an application under the Electricity Act is successful, a Direction is made by the Secretary of State that planning permission may be deemed to be granted under Section 40(1) of the Town and Country Planning Act 1971 and its 1990 successor. This Direction may include specific conditions and undertakings with regard to environmental issues, such as contamination of water supply, noise from construction and operation, archaeology, landscaping and creation of wildlife habitats. Planning Policy Guidelines for Renewables have been issued by the DoE/WO (1991c).

The analysis of the environmental implications of proposals for the construction and operation of a power station is required under the Electricity and Pipeline Works (Assessment of Environmental Effects) Regulations 1990. These regulations set out the information which should be included in an ES. The Electricity Act 1989 requires that:

'In formulating any relevant proposals, a licence holder or a person authorised by exemption to generate or supply electricity:

a) shall have regard to the desirability of preserving natural beauty, of conserving flora, fauna and geological or physiographical features of special interest and of protecting sites, buildings and objects of architectural, historic and archaeological interest; and

b) shall do what he reasonably can to mitigate any effect which the proposals would have on the natural beauty of the countryside or on any such flora, fauna, features, sites, buildings or objects.'

One aim of the EA is to demonstrate that these requirements have been met. Under Schedule 9 of the Electricity Act 1989, electricity utilities are required to make a statement regarding conservation and amenity, and undertake to consult with the appropriate national bodies.

Other legislation that bears on the environmental statement includes that governing releases to air and water, also dredging and solid waste disposal (see Sections 7.4−7.6). The Water Act 1989, the Environmental Protection Act 1990 and the Water Resources Act 1991 are particularly pertinent.

Where a proposed project impinges upon an area of conservation significance, consultation is required with the national body responsible. It is advisable to consult over nearby areas, even if no direct impact occurs. Organisations that may need to be approached in connection with the development of power stations are listed in Table 7.1.

Table 7.1 Possible consultees in the EA process (indicative only)

	Environmental issue	Activity	Consultees
1.	Land transport	Road/rail movements	LPA, LA, CC, DoT, B Rail
2.	Other infrastructure	Electricity, gas, water, shipping, aircraft	Relevant plc, CAA, DTI, Trinity House
3.	Land use	Farming, forestry, other industrial	LPA, MAFF, CC, LA
4.	Heritage	Archaeology, architecture	EH, DoE, relevant trusts
5.	Ecosystems	Habitats of flora and fauna (inc. birds/fish/man)	LPA, NRA, EN, MAFF, Sea Fisheries Comm.
6.	Drainage and flooding	Groundwater, river and tidal flows, storm surges	LA, LPA, NRA, MAFF, DoE
7.	Materials	Dredging, solid waste disposal, offshore works	MAFF, LA, EN, NRA, Sea Fisheries Comm.
8.	Pollution	Aquatic, atmospheric, terrestrial discharges	LPA, LA, HMIP, EN, NRA, CC, MAFF
9.	Amenity and leisure	Footpaths, beaches, general recreation	LPA, LA, NRA, CC, EN, Port Auth's
10.	Other projected developments	County structure plans and proposals yet to be made	LPA, DoE

Key:
DoE	Department of the Environment	LPA	Local Planning Authority
DTI	Department of Trade and Industry	LA	Local Authority
DoT	Department of Transport	CC	Countryside Commission
DEmp	Department of Employment	EN	English Nature
MAFF	Ministry of Agriculture, Fisheries and Food	EH	English Heritage
MoD	Ministry of Defence	CAA	Civil Aviation Authority
HMIP	Her Majestey's Inspector of Pollution	B Rail	British Rail
NRA	National Rivers Authority		

Note:
1. Where an English agency is quoted, this includes the corresponding bodies for N. Ireland, Scotland and Wales.
Not all organisations that need to be consulted have been listed (see also Table 2.2).

7.2.2 EA process and documentation

Environmental Assessment is an iterative process involving the appraisal of various project layouts and design criteria, and their adjustment as appropriate. The process is given in simple outline in Box 7.1. It embraces a range of environmental topics including socio-economics, landscape, health, wildlife conservation and amenity issues.

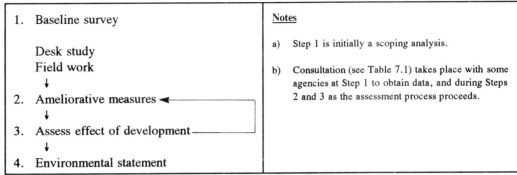

Box 7.1 EA process (construction and operation)

There are two phases in a power project where interactions may occur with the aquatic environment: during construction and during operation. No matter what type of power station is concerned, the processes and the type of consequential effects arising during construction tend to be broadly similar, varying in scale according to the plant itself and its site (see Chapter 4 on coastal engineering works). Impacts result from:

- the change in land use
- the source of materials
- the transport of materials
- the management of materials on site
- the management of wastes.

Assessment is largely a combination of the identification of the quantities and nature of materials involved, the transport routes to be used, the treatment processes to be used on site, the nature and fate of the wastes, and the interaction between these activities and the local environment. Loss or disturbance of habitat is the most obvious effect of construction activities and the significance of the loss has to be identified.

During operation the effects become more specific, some being peculiar to the energy sector.

The principles and procedures outlined here have mainly been developed from fossil and nuclear plant, and to a lesser extent from windfarms. They are equally applicable to new generating systems including 'renewable' energy devices such as solar energy and small combustion plant systems. All that changes is the emphasis given in the analytical appraisal, which depends on:

- effects peculiar to the technology, e.g. electromagnetic interference effects caused by the rotation of wind turbine blades

- the size of the development. A traditional large (2000 MW) fossil power station might occupy over 40 hectares (ha) on completion, with a wider area being affected during the construction phase. Stations powered by gas tend to be more compact, e.g. 15 ha for 1500 MW, because there is no need for large fuel stockpile areas as with a coal-fired plant. Gas-fired power stations may be built in different sizes from basic units of about 340 MW. Individual renewable energy systems are much smaller (e.g. 0.2–10 MW) but are spread out over a large area, e.g. 300 ha for an 8 MW windfarm using 250–300 kW generating units.

The significance of an impact is established by comparison, where appropriate, with existing background conditions as well as with environmental quality criteria in the form of regulations, standards and guidelines.

It is important that comparable standards of assessment are applied to all forms of electricity generation and experience suggests that this is feasible.

The end products of the environmental assessment process may include:

- an Environmental Statement
- supporting study reports
- a non-technical summary.

This documentation will be available for public consideration. The amount of documentation produced for smaller electricity projects will reflect the nature of the project and the environmental sensitivity of the affected area.

7.2.3 Environmental Statement

This is a technical document that is the end product of the assessment process. It aims to give a realistic picture of the project, the way the project is expected to impinge on the environment and the significance of these effects. An ES is a précis of the significant findings of an EA. Typical chapter headings for an ES are illustrated in Table 7.2 for thermal and wind systems. The amount and type of information presented will vary with the size, type and location of the generating scheme.

Table 7.2 Examples of ES chapter headings

Thermal stations	Wind energy projects
Chapter 1 Non-technical summary	Chapter 1 Introduction
	Chapter 2 Wind power in England and Wales
Project description	
Chapter 2 The site	**Project description**
Chapter 3 The development	Chapter 3 The site
Chapter 4 Constructing the station	Chapter 4 The wind park and its operation
	Chapter 5 Constructing the wind park
Environmental analysis	
Chapter 5 Atmospheric emissions	**Environmental assessment**
Chapter 6 Noise and vibration	Chapter 6 Landscape and visual
Chapter 7 Terrestrial ecology	Chapter 7 Flora and fauna
Chapter 8 Aquatic ecology and water quality	Chapter 8 Property and agriculture
Chapter 9 Coastal geomorphology	Chapter 9 Noise
Chapter 10 Landscape and visual effects	Chapter 10 Television and radio interference
Chapter 11 Employment	
Chapter 12 Transport	**Summary**
Chapter 13 Safety	Chapter 11 Summary
Glossary/abbreviations	Glossary/abbreviations
Appendix 1	**Appendices**
Legislative requirements	
Appendix 2	
Organisations consulted during preliminary site studies	

The ES should address the concerns of specialist staff employed by the regulatory and planning authorities. It will be advantageous to discuss chapters of the ES with regulatory bodies as the analysis develops and at the draft stages.

The environmental analysis should include a review of the existing environmental circumstances at the proposed site and its environs. It should describe and assess the most significant effects arising from construction and operation and should explain how these effects are to be controlled and, where necessary, mitigated.

Table 7.3 summarises the activities associated with each phase in the life of various types of power station. The extents of their effects depend on the type and size of a project as well as on its location, detailed design and method of operation. In most cases these activities are well understood, not least because most are common to other types of industrial and commercial development and are not exclusive to power stations.

The size and complexity of a power station make it necessary to demonstrate good environmental performance for a variety of ancillary activities such as fuel and waste transport, and support facilities such as back-up generator systems, incinerators and sewage works. It may also be necessary to address issues raised in other chapters in this publication, e.g. Chapter 6 on linear development (for gas pipelines and transmission lines) and Chapter 4 on river and coastal engineering (for the construction of cooling water intakes and outfalls).

7.2.4 Supporting reports

These may be prepared by consultants, in-house specialists, or both in conjunction. The corresponding chapters in the ES will be a précis of the supporting papers. These papers will identify the most significant environmental impacts of the proposed project and their importance measured against criteria appropriate to the discipline. They will also discuss measures that have been included in the design to avoid or reduce potential impacts throughout construction and operation.

Table 7.3 Aspects associated with potentially significant impacts (broad indication only)

		Construction	Operation	Decommissioning
Site	- scale	■		
	- access by land	■	■	■
	- access by water[1]	■	■	■
	- capital works on land	■		■
	- capital works in water[1]	■		■
	- maintenance dredging	■	■	■
	- landscape/visual	■	■	■
	- restoration	■		■
	- infrastructure		■	
Ancillary	- pipeline	■		
	- grid connection	■		
Raw materials	- demand	■	■	■
	- source	■	■	■
	- transport	■	■	■
	- storage	■	■	
Routine discharges	- gases		■	■
	- liquids	■	■	■
Waste disposal	- solids	■	■	■
Site activities	- noise/dust/odour	■	■	■
	- lighting	■	■	■
Workforce	- demand	■	■	■
	- source	■	■	■
	- housing	■		
	- transport	■	■	■
	- support	■	■	■
Regional	- infrastructure	■		
	- employment	■	■	
	- economics	■	■	
Safety/hazards		■	■	■

Note: [1] e.g. in rivers, estuaries or at sea

7.2.5 Non-technical summary

This will be much shorter than the full ES. It will explain to the interested layman:

- the nature and appearance of the project
- its impact on the environment
- the sorts of Consents that have to be obtained before it can be constructed.

7.3 INTEGRATED POLLUTION CONTROL (IPC)

Pollution control in the UK is moving towards an integrated approach embracing impacts with air, water and land, under the direction of HMIP. The main objectives of IPC are (DoE/WO, 1991b):

- to prevent or minimise the release of prescribed substances and to render harmless any such substances when released

- to develop an approach to pollution control which considers discharges from industrial processes to all media in the context of the environment as a whole.

At present, the most closely integrated assessment procedure is for radiological discharges from nuclear installations (see Box 7.2). The methodology for IPC of conventional pollutants is under development.

Box 7.2 Exposure to radiological discharges

> Impacts in terms of radiological exposures via air, water and solid waste management are separately modelled and then summed to give exposures to:
>
> - 'critical' local groups or individuals, i.e. those potentially likely to experience highest exposures
>
> - the total UK population ('collective' exposure).
>
> In each case, discharges are limited under the Radioactive Substances Act 1960, so that the total exposure is within authorised bounds. These are set out in the Ionising Radiation Regulations (1985), and take account of an EC Directive on radiation protection and the recommendations of the International Commission on Radiological Protection (ICRP). As well as setting limits, there is a legal requirement to reduce exposure to levels as low as reasonably practicable (ALARP). Total exposure includes:
>
> - exposure from discharges to air
> - plus exposure from discharges to water
> - plus exposure from waste transport
> - plus any exposure from direct radiation from the site
> - plus radiation due to the transport of irradiated fuel.
>
> Discharges from any adjacent power stations are taken into account, together with natural background radiation levels, when judging the significance of the additional radiation exposure. An ES should outline the assessment process and results, though not all of the detailed data would need to be given.
>
> 'Critical' exposure groups are identified in 'habit surveys' which reflect diet and lifestyles and are carried out by the Ministry of Agriculture, Fisheries and Food.

7.4 AIR QUALITY

7.4.1 Existing environment

Air quality impacts can be considered at three levels

- at a local level, where short- or long-term impacts on health, nature and the environment are of prime interest

- regional and trans-boundary levels, acid deposition being a prime example

- global level, where pollution effects can affect the earth's climate.

Some of these issues are described in more detail in Box 7.3.

Box 7.3 Some air quality issues

> *Greenhouse gases*
> The so-called greenhouse gases implicated in global warming and possible climate change include carbon dioxide (CO_2), methane (CH_4) and nitrous oxide (N_2O). CH_4 is a more potent greenhouse gas than CO_2, but the latter receives most attention because of the large quantities emitted. Atmospheric concentrations of CO_2 are known to have increased since pre-industrial times from about 280 ppm (by volume) to about 350 ppm in 1990.
>
> *Atmospheric CO_2*
> Atmospheric CO_2 concentrations could double before the middle of next century compared with levels around 1870 (IPCC, 1992), owing to continued industrial expansion, population growth and increasing energy demand. N_2O, which has a long atmospheric residence time, is produced by burning fossil fuels including biomass. Atmospheric concentrations of N_2O have increased from a pre-industrial value of 280 to 300 ppb (by volume) and have risen annually by close to 0.4% for the past decade (NSCA, 1991).
>
> *Ground-level ozone*
> Ground-level ozone is a threat to health because of its effects on respiratory performance. Concentrations have been subject to survey in recent years by WSL and are reviewed by the UK Photochemical Oxidants Review Group of the DoE (UKPORG, 1987). Nitrogen oxides are implicated in ozone production (UKPORG, 1990).
>
> *Acid deposition*
> Acid deposition is a regional-scale, trans-boundary issue. Existing soil and surface water acidity levels influence how much extra acid deposition can be tolerated. This has implications for emissions of SO_2, hydrogen chloride (HCl) and NO_x. Data are being reviewed on a European scale with regard to 'critical loads', i.e. acid deposition levels, at the threshold of change where damage occurs (Roberts *et al.*, 1992). The critical load for a soil depends on the dominant soil type allowing for factors such as mineralogy, texture and land use. Provisional maps for UK soils have been presented by DoE (1991a).
>
> *Stratospheric ozone*
> Another global issue is the condition of the ozone-rich layer in the stratosphere some 15–50 km above the earth's surface. Ozone is produced naturally at that level by the action of the ultraviolet radiation in sunlight on oxygen. Since the late 1970s, the layer above the Antarctic has thinned each spring (i.e. October), sometimes by over 50%, and thinning is now reported over Europe. The reason for this is the effect of chlorine derived mainly from long-lived man-made chlorofluorocarbons (CFCs) and halons, which destroy ozone. The position has been summarised by a UK Stratospheric Ozone Research Group (UKSORG, 1990, 1991). The ozone layer acts as a sun screen and its thinning increases the risk of sunburn, skin cancers and eye cataracts.

Data on the prevailing regime are available from the DTI's Warren Spring Laboratory (WSL) (planned to move to Harwell). Atmospheric pollutants of prime interest, because of potential short-term or long-term effects on health, are sulphur dioxide, suspended particulates (smoke), oxides of nitrogen (NO_x) and lead. Results of surveys are published by WSL, and trends from year to year are considered. In some cases it may be possible to get additional information on local background levels from local authority environmental health departments.

A variety of toxic trace metals, including lead, and trace gases are also monitored by WSL.

Information on trends in air quality conditions including smoke and sulphur dioxide (SO_2) concentrations are given in the DoE's annual *Digest of Environmental Protection and Water Statistics* (e.g. GSS, 1990). Reviews of the UK natural and man-made radiological environment are carried out by the National Radiological Protection Board, which also issues general discussion documents intended for wide readership (e.g. NRPB, 1981). Sources of exposure include radon gas leaking from soils and rocks and accumulating in buildings. Information on background conditions for specific pollutants, e.g. nitrogen oxides (NO_x), can be supplemented by local check measurements. It may not always be practicable for various reasons to carry out an extended measurement programme locally.

Background concentrations of pollutants tend to be higher in conurbations owing to the concentration of traffic, domestic and commercial activity and the low heights of the discharge of pollutants from these activities.

UK air pollution is monitored at numerous sites (Table 7.4), including 17 automated sites recording ozone and other pollutants.

Table 7.4 Air quality monitoring sites[1]: December 1990

Pollutant	No. of sites[2]
Ozone[3]	17
Carbon monoxide[3]	5
Nitrogen oxides[3]	12
Acid deposition, nitrogen dioxide, sulphur dioxide and ammonia	32
Atmospheric hydrocarbon	2
Trace gases	1
PAN, hydrogen peroxide, hydrocarbons and aerosols	2
Smoke and sulphur dioxide sites	261
Other sulphur dioxide sites[3]	9
Lead	18
Multi-element sites including lead	5

Source: Warren Spring Laboratory.
Department of Trade & Industry

Notes:
[1] Monitored on behalf of the Department of the Environment by Warren Spring Laboratory, Department of Trade and Industry; Harwell Laboratory; University of Manchester Institute of Science and Technology and the Institute of Terrestrial Ecology.
[2] Monitoring sites for different air pollutants may be located in the same place.
[3] Automated monitoring sites.

7.4.2 Air quality regulations and standards

Standards applicable to England, Scotland and Wales are set by the Air Quality Standards Regulation 1989 for various pollutants. The Regulations implement EC Directives setting air quality standards for sulphur dioxide and suspended particulates (80/779), nitrogen oxides (85/203) and lead in air (82/884). The Air Quality Standards Regulations (Northern Ireland) 1990 apply similar provisions. Data are given as mandatory limit values and guide values which it is desirable to achieve. Standards are set on the basis of medical studies and toxicological testing.

The World Health Organisation publish additional information and criteria to include protection of vegetation. Selected air quality criteria are given in Table 7.5 for N_2O, SO_2 and lead in air. HMIP has published a technical guidance note on dispersion (HMIP 1993). This is one component of the Integrated Pollution Control system. The Environmental Protection (Prescribed Processes and Substances) Regulations 1991 indicate which processes are regulated and whether they are dealt with by HMIP or by local authorities. Substances to be controlled in relation to air pollution are given in a schedule to these regulations (plus similar information for releases to land and water).

The concept of 'critical target loads' has been developed to account for the effects of acidifying species on different soils and watercourses and to protect sensitive species, communities and ecosystems (DoE, 1991a; Roberts et al., 1992). Known critical loads for sensitive areas help to define target loads, which are still being determined, as the permitted pollutant deposition load for an area. Depending on circumstances including social pressures and economic constraints, target loads may be lower or higher than critical loads.

Table 7.5 EC and WHO air quality limit and guideline values (in $\mu g/m^3$, standardised at 293 K and 101.3 kPa pressure)[d]

	Nitrogen dioxide	Sulphur dioxide	Lead
Human health			
EC limit value			
98th percentile	200[a]	350[b]	
50th percentile	-	120[b]	
annual mean			2
EC guide value			
98th percentile	135[a]		
50th percentile	50[a]		
annual mean	-	40-60	
WHO guideline			
1-hour mean	400		
Vegetation			
WHO guideline			
annual mean	30[c]	30	

Notes:
    a. Expressed as a percentile of hourly mean values throughout a year.
    b. Expressed as a percentile of daily mean values throughout a year. These limits apply when smoke levels are less than $150\mu g/m^3$ as a 98 percentile of daily mean values and $40\mu g/m^3$ as a 50 percentile of daily mean values. Otherwise stringent criteria apply.
    c. Applied in combination with annual mean sulphur dioxide less than $30\mu g/m^3$ and annual mean ozone less than $60\mu g/m^3$.
    d. See also NSCA (1992 and annually).

EC Directive 89/609/EC on the limitation of emissions of certain pollutants from large combustion plants, defines ceilings and reduction targets for emissions of SO_2 and NO_x from existing plants in the UK and other EC countries. These provisions will influence trans-boundary pollution as well as national deposition levels for these pollutants. Long range trans-frontier pollution was previously the subject of the 1985 Protocol of the UN Commission for Europe's Convention on Long-Range Trans-boundary Pollution, though this was not ratified by either the UK or EEC.

EC Directive 89/609/EEC also proposes limits on SO_2, NO_x and dust emissions for new solid- or liquid-fuelled combustion plant.

Trans-boundary effects of radiological discharges are subject to assessment under Article 37 of the Euratom Treaty.

In 1987, international agreement on substances that deplete the ozone layer was reached (the Montreal Protocol) to protect the stratosphere ozone layer by reducing the production and use of CFCs, halons and carbon tetrachloride by various targets up to the year 2000 and then was strengthened in 1990 by agreement to eliminate them by that date. This was endorsed by EC/373/28, 28 December 1991. All industries including the power industry are affected by the need to replace CFC-based refrigerants, blowing agents for insulation, solvents and aerosol propellants.

7.4.3 Impact evaluation

Effects of air quality are controlled by limiting emissions at source or by releasing materials via sufficiently tall stacks and encouraging plume rise, so that by the time the contaminants reach ground level they have been diluted by atmospheric turbulence or wind shear. The degree of dilution is usually calculated for a variety of meteorological conditions so that incremental changes in the air we breathe can be shown to be acceptable, or otherwise, relative to air quality standards and/or existing background levels. In some cases, pollutants undergo chemical changes in the atmosphere which modify their significance with regard to health considerations, e.g. nitrogen oxide conversion to the more hazardous nitrogen dioxide, or NO_x interacting photochemically in air to form ozone in the presence of volatile organic compounds (VOCs).

Dispersion calculation models require information on emission rates. An example of emission data for a large coal-fired station is given in Table 7.6. Efflux velocity and temperature may also have to be specified. Particulates emitted after filtration are small and tend to disperse as gases. They contain a variety of trace elements including radioactive isotopes originating in coal (Lim, 1979).

Table 7.6 Typical emission rates for 3x660 MW coal-fired station (without FGD)

	Concentration in flue gas	kg/ms @ MCR	kg/year @ 56% L.F.
SO_2	1200 vpm	6.9	120.0
NO_x	550 vpm	1.5	26.0
Particulates	115 mg/m³	0.22	3.9
Coal consumption		215	3,800.0

Assumptions: Calorific value 24,930 kJ/kg
Sulphur content 1.6%
Ash content 16.4%
Precipitator efficiency 99.3%

Box 7.4 Frequency and spatial distribution of pollutants

The consequence of the variability of air quality due to turbulence and other factors is illustrated in Figure 7.1 based on a two-year survey around a 2000 MW coal-fired station with a 198-m-high multi-flue chimney. Off-axis background measurements were subtracted from measurements made when the wind was blowing from the power station, to generate a statistical picture of ground-level concentrations at the range of peak impact, several km away.

Figure 7.1 gives the frequency with which SO_2 concentrations were exceeded for selected averaging times from about 3 minutes to 1 year. The curves are normalised to represent the effect of the power station at a point which receives an average proportion of winds blowing from the station. It can be seen that these ground-level exposures vary significantly with averaging period, e.g. an hourly average of 20 $\mu g/m^3$ is exceeded between 4% and 5% of the time, whereas the annual average is around 2–3 $\mu g/m^3$. Current computer models calculate similar information depending on the quality of input data.

For comparison with health criteria, daily averages of ground-level concentrations are calculated for SO_2 and NO_x. One-hour average data are of interest for NO_x. Examples of the output from such calculations, using a standard computer modelling package for plume rise and atmospheric dispersion, are given in Figure 7.2 for a stand-by generator. These site specific results are presented as short-term peak concentrations versus distance for different atmospheric stability conditions which govern rate of dilution. Calculations of exposures allowing for long-term windspeed and duration climatology are presented in the form of isopleths (contour maps) to illustrate areal distribution (Figure 7.3).

Fig. 7.1

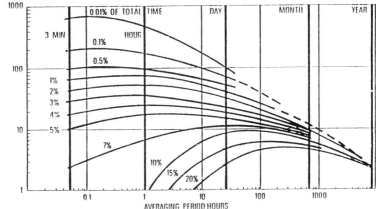

Fig. 7.2

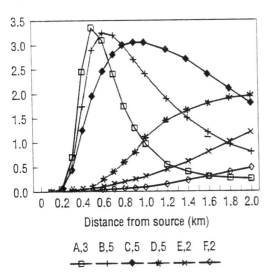

Fig. 7.3

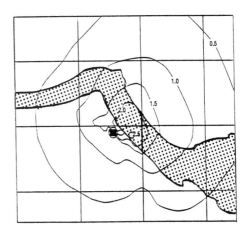

Figure 7.1 *Frequency distribution of SO_2 concentrations from 2000 MW power station at radius of maximum effect (~8 km) $\mu g/m^3$*

Figure 7.2 *Peak hourly average GLCs for 1 g/sec emission rate (site specific results) results for six stability class/windspeed (m/sec) configurations*

Figure 7.3 *Estimated annual average ground-level concentrations of NO_x (NO + NO_2) $\mu g/m^3$ CCGT power station site*

Predictions of air quality are generally based on mathematical methods and procedures, tested against field dispersion trials data collected since the 1930s.

Assessing the significance of a change in air quality can be complex, especially when the short-term peak exposure is the criterion of interest for health protection. This is the case, for example, with NO_x where the 98th percentile of 1-hour average exposure is of interest. Complications arise because both the background concentrations and the contribution of an elevated point source, such as a power station stack, may vary significantly on quite short timescales at a given place. This is because of the variability of atmospheric turbulence over timescales of seconds to minutes, combined with longer-term swings in average windspeed and wind direction (see Box 7.4).

As far as possible, models have been refined to take account of variations due to flow over complex terrain, the proximity of other buildings and the washout or dry deposition of contaminants from atmospheric suspension. A number of such models are on the market but only those acceptable to the regulators (i.e. HMIP or the local authorities) should be employed. Numerous environmental consultants offer modelling capabilities, often based on the Industrial Source Complex Model prepared and updated by the US Environmental Protection Agency.

Regional-scale emissions are now subject to assessment and control. Trans-boundary acidic pollution has gained greater interest during the last two decades because of effects in areas sensitive to acid deposition. EC legislation has produced requirements to reduce discharges of SO_2 and NO_x on a national scale and this has implications for individual plant designs, e.g. fitting desulphurisation and low NO_x emission systems. Radiological emissions are subject to assessment of effects in neighbouring countries under Article 37 of the Euratom Treaty. Episodes of high ozone concentration depend on regional emissions of NO_x and hydrocarbons and prevailing weather conditions. The build-up of ozone in the lower atmosphere is favoured by stable, sunny conditions.

Power plant operation contributes to greenhouse gas emissions, mainly via carbon dioxide from the burning of fossil fuels or methane from gas leakage and coal mines. Carbon dioxide emissions per kWh vary from negligible levels to approximately 1 kg CO_2/kWh according to the generation system employed. Listed from lowest to highest these are:

- nuclear and renewables
- combined gas cycle turbine
- oil-fired generation
- coal-fired generation.

An Intergovernmental Panel on Climate Change has studied global warming mechanisms (IPCC, 1992) and potential consequences (IPCC, 1990). Implications for the UK were considered by the Climate Impacts Review Group (UKCCIRG, 1991).

While peak ground-level concentrations are of prime importance in air quality assessment, concentrations at a greater distance will be of interest if a particular pollution-sensitive ecological site or species is exposed, such as lichens sensitive to sulphur dioxide.

Apart from source emission terms, dependent on engineering design and patterns of plant use, the main data input to the modelling process is meteorological. If, as may well be the case, there is a shortage of long-run data, the data from the nearest standard meteorological station have to be adopted. In the UK, the Meteorological Office can supply data, analysed by frequency of occurrence, for standard wind speed, wind direction and atmospheric stability bands. Atmospheric stability is a measure of the dispersive power of the atmosphere and particularly depends on the vertical profile of temperature.

Table 7.7 Summary of best estimate airborne discharges for a PWR reactor

Nuclide	Discharge (Bq/yr)	Nuclide	Discharge (Bq/yr)
H-3	2.1×10^{12}	I-130	1.9×10^6
C-14	3.0×10^{11}	I-131	2.7×10^8
Ar-41	2.2×10^{11}	I-132	3.2×10^8
		I-133	8.9×10^8
Kr-83m	3.7×10^{10}	I-134	3.7×10^8
Kr-85	3.4×10^{12}	I-135	2.9×10^8
Kr-85M	7.8×10^{11}		
Kr-87	1.0×10^{10}		
Kr-88	1.4×10^{11}	Total particulate	1.9×10^7
Kr-89	3.7×10^8		
		Total actinides	1.4×10^8
Xe-131m	1.9×10^{12}		
Xe-133	1.5×10^{14}		
Xe-133m	1.6×10^{12}		
Xe-135	9.3×10^{11}		
Xe-135m	5.2×10^{12}		
Xe-137	1.0×10^{10}		
Xe-138	3.0×10^{11}		

Table 7.8 Examples of exposure levels

Annual individual dose (μ Sv/yr)		Collective dose to population of the United Kingdom (man-Sv/yr)	
To hypothetical most exposed individual due to PWR in normal operation	16.7	PWR operation	0.08
To average individual in everyday life from natural sources	1870 (1070 excluding radon)	Natural background	100,000

Table 7.9 Estimated maximum individual risk of death consequent upon PWR operation

	Early death (per yr)[1]	Fatal cancer (per yr)[2]
Normal operation[3]	0	1 in 6,000,000 (1.7×10^7)
Accidents:		
Design basis[4]	0	1 in 10,000,000,000 (8.5×10^{11})
Containment bypass[4]	1 in 100,000,000,000 (1×10^{-11})	1 in 500,000,000 (2×10^{-9})
Degraded core[4]	1 in 500,000,000 (2×10^{-9})	1 in 5,000,000,000 (2×10^{-10})
'Everyday' risk for comparison	1 in 3000	1 in 300

Notes: 1. This is the annual average risk of death in the UK from accidents of any kind.

2. This is the annual risk in the UK of developing a fatal cancer averaged over a person's lifetime.

3. These figures apply to a hypothetical individual who is assumed to have the habits of each of the critical groups.

4. These figures are those for an individual assumed to live continuously at a distance of 1 km from the reactor.

For radiological exposures, the combined effects of exposure along all pathways, i.e. air, water and food, are computed and evaluated for carefully selected critical groups in the local community. Hence these have to be modelled and summed. Exposure routes evaluated for emission to the atmosphere are:

- direct radiation from the plume
- inhalation of radioactive gases and particulates
- irradiation from deposited materials.
- ingestion of radioactivity taken up by crops or by secondary routes such as intake by cows and transfer to milk.

To do this, atmospheric dispersion deposition and radiological pathways to man are modelled. The nature and quantities of emitted radioactive materials vary with the type of power station. An example of estimated atmospheric discharges for a large pressurised water reactor (PWR) is given in Table 7.7. Particulates from coal-fired power stations contain trace quantities of natural radioisotopes which can also be assessed radiologically.

Examples of the outcome of site-specific calculations of the highest levels of exposure of local individuals and the collective UK dose are given in Table 7.8. For UK nuclear reactor systems, computer programs have been developed to evaluate source terms and exposures for routine discharges, allowing for the various potential environmental pathways, recognising dispersion, dilution and concentration processes.

For nuclear sites an assessment is also made of the probability and magnitude of accidental exposure of the public, taking account of engineering design measures to contain associated risks below acceptable levels.

Exposures due to routine discharges can be expressed in terms of risk and compared with risks normally experienced by society. An example of this is given in Table 7.9. All exposure pathways, not just air, are included in this assessment.

The significance of an impact is established by comparison with background levels and environmental quality criteria, in the form of regulations, standards or guidelines.

7.4.4 Mitigation measures

Measures which can be taken to reduce air pollution include:

- tall stacks and heated plumes to promote plume rise and facilitate dilution by atmospheric dispersion

- minimum discharge efflux velocities to avoid down-wash

- electrostatic precipitators or bag filters to contain particulate discharges within authorised emission limits

- flue gas desulphurisation

- burning low-sulphur fuels

- controlled combustion conditions, with adequate residence time for incinerators and waste-derived fuel systems

- NO_x emission controls, e.g. burner design, steam injection, regeneration systems

- good housekeeping including damping down to avoid fugitive dusts from coal stocks and at ash disposal/utilisation schemes

- liming catchments to offset acid deposition

- nuclear containment, shutdown and filtration systems

- barriers or ventilation to reduce natural radon gas accumulation in buildings

- gradual replacement of CFCs by suitable alternatives

- planting trees and building flood protection walls to counter global warming. Possible adaptation of agriculture in longer term.

7.5 WATER QUALITY

7.5.1 General

The traditional view of the seas was as an infinite supplier of food and infinite sink for wastes, while the function of rivers was for transport of goods or wastes to that infinite sink; this has largely passed. An intermediate approach has dominated over the last decade, that of assimilative capacity, but the sensitivity of the aquatic environment to exotic man-made compounds, even at hardly detectable levels, is leading towards the increasing application of the precautionary principle. In the UK a system of classification of controlled waters, absolute and use-related water quality objectives and standards is evolving as a means of management of the aquatic environment.

In environmental assessment it is advisable for proposals for discharges to be brought before the National Rivers Authority (NRA) or HMIP as appropriate at an early date. Similarly, the need for extraction from other than coastal waters needs to be discussed, again with the NRA, while both extraction from coastal waters and any offshore activities associated with construction need to be discussed with MAFF. Small inland water bodies and coastal margins may be designated SSSIs, because of the bird interest, saltmarsh, or the aquatic fauna; the great majority of estuarine sites possess multiple nature conservation designations, and such controls are increasing. In each of these cases it is appropriate to consult English Nature or its equivalents.

7.5.2 Water quality regulations and standards

The National Rivers Authority (NRA) is responsible for maintaining the quality of controlled waters, including rivers, lakes and reservoirs, estuaries, coastal waters to the 3-mile limit, and groundwater. It is increasingly moving towards the management of water quality by means of Statutory Water Quality Objectives (SWQOs), encompassing a set of Use Classes each having an accompanying set of Standards (aesthetic, chemical, and potentially biological) according to the needs of that Use. Further, a set of Environmental Quality Standards (EQSs) is continuing to develop for the most dangerous substances (heavy metals and chlorinated organic compounds), listed in the UK's Priority Substances List, and the international Lists I and II. It is the intent of the NRA (NRA, 1990) to move towards the ecotoxicological appraisal of complex effluents, thus protecting biological systems.

The NRA was established under the provisions of the Water Act 1989. Under the Water Resources Act 1991 it is for the Secretary of State to prescribe by Regulations a system for classifying the quality of controlled waters, and then to set Statutory Water Quality Objectives (SWQOs) by reference to that classification. A current classification scheme does exist, that of the National Water Council (NWC) incorporating an estuarial scheme developed by the Marine Pollution Monitoring Management Group (MPMMG), but has no statutory basis. The new classification scheme was to be established in 1992, with SWQOs being set on a rolling programme thereafter: it is the first such scheme to involve all of ground, surface, estuarine and coastal waters. Currently (7/92) the consultation phase over the new classification and SWQO scheme is ending, the Regulations to come before Parliament in Autumn 1992, intended to take effect before the end of the year (NRA, 1991a). It is important to realise that this is likely to be a continuing process, with new EQSs for specified substances, and objectives continually being introduced and old ones revised. The current revision of the guidelines issuing from the Oslo and Paris Commissions (OsParCom) is expected to emphasise the precautionary principle and sustainable use, requiring that all substances of anthropogenic origin be treated as dangerous until proven otherwise: this is likely to prove strong support to the implementation of NRA policy as described in the Kinnersley Report (NRA, 1990).

Fundamental to the establishment of the SWQO scheme and the WQSs for certain substances are international criteria and standards, primarily those established by the European Inland Fisheries Advisory Commission (EIFAC), which are advisory only, and the EC Directives, for which the UK Government has a statutory obligation. The pertinent EC Directives are listed in Table 7.10 and a provisional list of proposed EQSs related to EWQOs is given in Table 7.11. It is important to note that the regulator will tend to apply whatever criteria are available, whether draft, statutory or guideline, in setting consents or other licences. There is sometimes the opportunity to evolve appropriate criteria with the regulator (see also Chapter 5).

The NRA is also responsible for the management of freshwater fisheries, and, in order to accomplish this, sets by-laws controlling permits, seasons and minimum sizes. Similarly, the regionally based Sea Fisheries Committees (SFCs) are empowered to set similar by-laws for the coastal areas under their jurisdiction.

The Ministry of Agriculture, Fisheries and Food (MAFF) is responsible for the management of marine fisheries, their protection through the conservation of inshore (frequently nursery) populations under EC policy guidelines, the offshore disposal of solid waste, and liquid effluents arising beyond the 3-mile limit. As the sea disposal of industrial wastes is phased out towards 1995 (DoE, 1990b), it is expected that MAFF will turn more detailed attention towards other coastal operations, such as dredging and aggregate extraction: both these activities may occur when coastal-based power plant is constructed.

Both MAFF and NRA are responsible under the Department of the Environment for implementing EC Directives, consequent regulations, and other policy guidelines, such as those issuing from the Oslo and Paris Commissions, and the Ministerial Conferences on the North Sea (which the UK Government takes to include all UK coastal waters; DoE, 1990b).

The other authority likely to be encountered, especially where prescribed substances or processes are involved, is Her Majesty's Inspectorate of Pollution (HMIP). The Environment Protection Act 1990, in establishing a system of Integrated Pollution Control (IPC), described certain processes and substances as Prescribed. Under a Memorandum of Understanding between the NRA and HMIP, the latter takes the lead in managing Prescribed Processes, consulting with NRA as appropriate over the EQSs to be applied for discharges to controlled waters, whereas where no Prescribed Process but Prescribed Substances are involved, NRA take the lead, consulting with HMIP as appropriate. At no time will HMIP use less stringent standards than those currently utilised by NRA. Legislation is suggested which will combine HMIP with the NRA to form an Environment Agency (EA). The MAFF functions appear at this stage to be conserved.

In Scotland the analogous organisations to the NRA are the River Purification Boards (RPBs). HMIP is paralleled by Her Majesty's Industrial Pollution Inspectorate (HMIPI), and MAFF by the Scottish Office Agriculture and Fisheries Department (SOAFD). A Scottish Environmental Protection Agency is also imminent. In Northern Ireland, fisheries are controlled by the Department of Agriculture for Northern Ireland (DANI), and pollution control by the Department of the Environment for Northern Ireland (DoENI). NRA, MAFF and HMIP cover England and Wales only.

Table 7.10 Summary of EC Directives relevant to EA process for energy schemes

Date	Council Directive	Ref.
1975	Quality of surface water for drinking	75/440/EEC
1976	Pollution caused by the discharge of dangerous substances into the aquatic environment	76/464/EEC
1976	Bathing water quality	76/160/EEC
1978	Quality of fresh waters needed to support fish life	78/659/EEC
1979	Quality required for shellfish waters	79/923/EEC
1980	Protection of groundwater against pollution caused by certain dangerous substances	80/68/EEC
1980	Quality of water intended for human consumption	80/778/EEC
1983	Limit values and quality objectives for discharges of cadmium	83/513/EEC
1984	Limit values and quality objectives for discharge of hexachlorocyclohexane	84/491/EEC
1984	Limit values and quality objectives for mercury discharges by sector other than the chlor-alkali electrolysis industry	84/156/EEC
1986	Limit values and quality objectives for discharge of certain dangerous substances included in List I of the Annex to Directive 76/464/EEC (carbon tetrachloride, DDT and PCP)	86/280/EEC
1988	Amending Annex 2 to Directive 86/280/EEC (The Drins: Dieldrin; Aldrin; Endrin; Hexachlorobenzene; Hexachlorabutadiene; Chloroform)	88/347/EEC
1989	List II substances — the implementation of EC Directives on pollution caused by certain dangerous substances discharged to the aquatic environment	DoE Circular 7/89
1989	[Surface Waters (Dangerous Substances) (Classification) Regulations, 1989 Statutory Instrument No. 2286]	[SI 2286]
1989	[Surface Waters (Classification) Regulations, 1989, Statutory Instrument No. 1148]	[SI 1148]
1990	Amending Annex 2 to Directive 86/280/EEC (1,2 dichloroethane, trichloroethane, perchloroethane, trichlorobenzene)	90/415/EEC
1991	Urban waste water treatment	91/271/EEC

Table 7.11 Proposed (UK) use-related water quality standards

Use	Application	Parameters	Value	Use	Application	Parameters	Value
Basic amenity	RLEC	Colour Transparency Odour Oil Litter Foam Biological growth Dissolved oxygen	 >10% sat	Commercial harvesting of marine fish for human consumption	EC	List II substances and prescribed substances under the Environmental Protection Act 1990	
				Commercial harvesting of shellfish for human consumption	EC	List II substances and prescribed substances under the Environmental Protection Act 1990	
				Water contact activities	RLEC	Bacteriological criteria	
General ecosystem	RLEC	Ecological Quality Index from BMWP, ASPT for rivers No. of taxa		Abstraction for potable supply	RL	EC Surface Water Directive List II Substances	
Special ecosystem	RLEC	*		Industrial and agricultural abstraction	RL	'No deterioration'	
Salmonid fishery	RL	Dissolved oxygen BOD Un-ionised ammonia Total ammonia pH Temperature max Temperature diff List II substances	5 mg/l 5 mg/l 0.021 mgN/l 0.78 mgN/l 6–9 21.5 °C 1.5 °C	Irrigation	RL	pH Conductivity Chloride Sulphate Fluoride Arsenic Boron Cadmium Chromium Copper Iron Lead Molybdenum Nickel Selenium Vanadium Zinc	5.5–8.5 1500 μS/cm 100–900 mgCl/l 150–135 mgSO$_4$/l 1 mgF/l 0.04 mgAs/l 2.4 mgB/l 0.02 mgCd/l 2 mgCr/l 0.5 mgCu/l 1–2 mgFe/l 2 mgPb/l 0.03 mgMo/l 0.15 mgNi/l 0.02 mgSe/l 0.08 mgV/l 1 mgZn/l
Cyprinid fishery	RL	Dissolved oxygen BOD Un-ionised ammonia Total ammonia pH Temperature max Temperature diff	3 mg/l 8 mg/l 0.021 mgN/l 1.5 mgN/l 6–9 28 °C 3 °C	Livestock watering	RL	pH Conductivity Chloride Sulphate Fluoride Dissolved oxygen Arsenic Chromium Copper Lead Nickel Zinc	6–9 3000 μS/cm 1000 mgCl/l 250 mgSO$_4$/l 2 mgF/l 3 mgO$_2$/l 0.2 mgAs/l 1 mgCr/l 0.2 mgCu/l 0.1 mgPb/l 1.0 mgNi/l 25 mgZn/l
Migratory fishery	RLEC	Dissolved oxygen BOD Un-ionised ammonia Total ammonia pH Temperature max Temperature diff List II substances	3 mg/l 6 mg/l 0.021 mgN/l 1.5 mgN/l 6–9 21.5 °C 1.5 °C				

R: rivers   L: lacustrine   E: estuaries   C: coasts

* Standards to be identified and applied according to the special requirements of identified sites.

7.5.3 Aquatic impact evaluation

Impacts on the aquatic environment arise from:

- the entrainment of water for cooling or dilution
- alterations to local tides or river flows
- the use of biocides to prevent fouling by the growth of aquatic organisms in the cooling water circuit
- leakage from recirculating systems
- the need to disperse liquid wastes
- the need to dispose of solid wastes
- the need to dispose of waste heat.

The first stage in the EA for the operational phase is to identify and quantify operational processes, then identify these elements of the local ecology at risk. For instance, the entrainment of cooling water by a river or coastal thermal station will normally result in the impingement of fish on screens, the entrainment through the system of smaller organisms, and the discharge of a thermal plume containing residual levels of any biocide used in anti-fouling the cooling circuits and other potentially toxic effluents such as those from flue gas desulphurisation (Table 7.12). The effects of these processes will depend upon their scale, the impact of taking in water, the precise locations and designs of intake and outfall structures, and the zoogeographic context. The effects of 'thermal pollution' may also require some attention, although this is rarely a problem for estuarine or coastal direct-cooled stations in temperate waters. A hydro, tidal or thermal plant may harm entrained animals by pressure, abrasion, shear and turbine or pump blade impingement: the actual damage caused will be consequent upon design, operating conditions, the volume of throughput, and the behaviour, size and number of organisms.

A variety of methods have been developed to calculate potential impacts. Expert computer systems may be employed to quantify the likely fish impingement of new coastal and estuarine stations in the UK. Species may be of commercial or nature conservation interest. Table 7.13 presents a comparison of fish impingement at a power station complex with commercial fish catches in the surrounding sea fisheries area. It can be seen that screen catches are not significant. A refinement is the use of the 'adult equivalent yield' approach: as coastal and estuarine plant tend to catch juvenile fish, the expected survival and growth rates of the species involved can be identified, together with a knowledge of the commercial fishing mortality and stock size. This allows the 'adult equivalent' in the fishery of the juveniles caught by the station to be estimated: thus the actual catch of the station is set in its proper context. In fresh waters, with more confined fish stocks, this is a straightforward procedure.

The study of the bottom-dwelling fauna (the benthos) during such an assessment may often be valuable in that their specific composition and abundance integrate local environmental conditions with time, whether hydrographic, water quality or sedimentological. These populations have a secondary direct importance, as food for birds on tidal margins, and food for fish; in some situations they may provide the bulk of the local primary and secondary productivity. They are also integrally involved in sediment/water contaminant transfer processes and often too in the consolidation of the sediments. The sensitivity to incremental change and disturbance stress of these populations is understood, and the effects of a development upon them may reasonably be predicted.

To investigate entrainment of small organisms through a power station cooling circuit it is pertinent, for a large-volume direct-cooled system, either to refer to analogous situations regarding species and intake design where field determinations of actual mortality and sub-lethal effects have been quantified, or to estimate the survival of the organisms involved by experiment, e.g. on fish lavae. It is important to have access to national databases held by MAFF and others, such as those on young fish surveys. It may become pertinent to collect information on the incidence of fish larvae locally. In each circumstance, loss has to be

Table 7.12 Effect of treated FGD effluent on a typical power station cooling water discharge

Contaminant	Increase in cw concentration µg/l	
	Once through	**Recirculated**
Chloride	13.0 mg/l	399.1
Nitrate	0.26 mg/l	8.09
Arsenic	0.05	1.46
Cadmium	0.01	0.29
Chromium	0.05	1.46
Copper	0.02	0.73
Lead	0.01	0.29
Mercury	0.005	0.15
Nickel	0.01	0.29
Zinc	0.02	0.73
Aluminium	0.24	7.32
Antimony	0.24	7.32
Boron	31.21	961.60
Fluoride	9.50	292.65
Iron	0.24	7.32
Manganese	0.48	14.63
Molybdenum	0.24	7.32
Selenium	0.05	1.46
Tin	0.24	7.32
Vanadium	0.24	7.32

Effluent flow rate = 23.8 l/s; CW discharge once through = 55 cumecs; recirculated = 1.6 cumecs
Source: Taylor *et al.*, 1988

Table 7.13 Comparison of published commercial fish landings for ICES sub-area VIIf (Bristol Channel) with annual screen catches from Hinkley Point A and B power stations estimated from monthly samples

	United Kingdom commercial landings (tonnes)			Power station (tonnes)
Year	**1980[1]**	**1981[2]**	**1982[3]**	**1981–84 annual mean**
Species				
Demersal				
Cod	97	152	156	<0.1
Haddock	20	9	5	-
Plaice	215	197	173	<0.1
Redfish	-	-	1	-
Saithe	6	12	25	-
Skates and rays	439	451	470	<0.1
Soles	122	142	158	-
Whiting	126	106	73	3.8
Other demersal	1,558	2,575	2,665	0.1
Pelagic				
Herring	0	1	3	<0.1
Mackerel	19,612	10,555	16,069	-
Other pelagic	276	241	182	2.6

Notes:

1. HMSO (1981) *Sea Fisheries Statistical Tables*
2. HMSO (1982) *Sea Fisheries Statistical Tables*
3. HMSO (1983) *Sea Fisheries Statistical Tables*

assessed within the context of local populations, other sources of mortality, and the ecological, commercial and wildlife conservation significance of that loss.

Direct-cooled thermal and hydro plant on river or lake systems, and tidal plant on estuaries, may involve the rerouteing and constraining of the flow on a significant scale, and consequently significant effects on water levels, tidal regimes, dispersive characteristics and sediment transport. These aspects require assessment. Cooling towers or spray cooling may be needed if water supplies are insufficient for direct cooling. An incidental effect of aeration within indirect systems such as these is that the effluent can have significantly higher oxygen levels than the receiving water.

Simulation of water movement, either by physical models or mathematical means, provides a useful way of predicting the likely scale of interaction of a scheme, whether upon the hydrography, sediment behaviour or transport of contaminants. It is important to approach modelling from an established and robust empirical understanding of local dispersion and sediment behaviour. Modelling should be on a scale appropriate to the scale of the processes involved; for example, full 3-D modelling would not normally be anticipated for indirect-cooled systems in temperate waters.

Simulation modelling of different levels of complexity may be useful at several stages in the EA process. Initial scoping studies using simple but robust models can aid in the early optimisation of plant design, while identifying critical limiting factors. Later, more sophisticated simulations based upon field hydrographic, sedimentological and water chemistry studies may allow more refined predictions. For thermal plant using direct cooling, the dispersion of heat may be simulated. This itself will provide the basis of contaminant dispersive predictions; for example, residual biocide levels can be predicted by 1D empirical pseudo-first-order decay models, using temperature as a conservative indicator of dispersion. Validation may be carried out by simulating and confirming the behaviour of other local discharges at the site location, or by the release of tracers, fluorescent dyes or spores simulating such discharges.

The disposal of liquid wastes, including biocides, by whatever means, is subject to the regulations described earlier. Investigative hydrographic studies and computer simulation of the effluent plume may become necessary, dependent upon scale and the nature of the wastes. For large cooling-water systems, station design optimisation requires hydrological modelling in order to minimise re-entrainment of heated water and loss of station efficiency, but it is helpful if the model formulations are arranged to allow environmental effects to be considered as well.

Operational effects are thus widely divergent, and the means of assessment equally so. A 'shopping' list of information would include, apart from the details of design:

- an initial scoping review of local ecological habitats

- a more detailed study of sessile river, sea-bottom or shore animals if they are likely to be affected

- a review of local fisheries

- a quantification of the fish likely to be impinged, their immediate and adult-equivalent value in the fishery and a wildlife conservation evaluation

- a quantification of the effects of entrainment of fish larvae, other zooplankton, and phytoplankton through the station itself or when in contact with the thermal effluent plume

- a knowledge of the behaviour of the plume, possibly up to the level of a validated 2-dimensional, layered, thermal simulation

- a knowledge of the levels, toxicity, behaviour and fate of the contaminants carried in that plume

- a knowledge of analogous situations elsewhere and the likely veracity of these assessments

- identified means of mitigation.

As discussed in Section 7.4, calculated radiological exposures from nuclear power stations via water are added to those from other routes to give collective exposure and critical group exposures. For the source term, the type (isotopic composition) and quantities of emissions to the aquatic environment are specified in detail, e.g. Table 7.14. An assessment is made of dispersion in the aquatic environment and radiological doses to people via various routes including consumption of seafoods. The assessment includes consequences for fishermen as well as people who consume fish. An example of the types of critical groups identified and considered in such an analysis is given in Table 7.15.

Table 7.14 Summary of liquid discharges for a PWR reactor (case study)

Nuclide groups	Dicharge (Bq/yr)
Total halogens	3.4×10^{11}
Tritium	2.1×10^{13}
Total caesiums and rubidiums	1.2×10^{11}
Total other fission and corrosion products	4.8×10^{11}
Total actinides	4.8×10^{9}

Table 7.15 Dietary and behavioural factors for critical groups identified for liquid effluent discharge assessments (case study)

Group 1	
Fish	50 kg/y
Crustacea	7.9 kg/y
Occupancy (contact with silt)	780 hr/y
Net handling (fishermen)	520 hr/y
Group 2	
Occupancy (contact with silt)	1027 hr/y
Net handling (fishermen)	514 hr/y
Group 3	
Net handling (fishermen)	936 hr/y

Note:

The doses to the critical group were calculated assuming that they had the habits of the group which gave the maximum dose (Group 1).

7.5.4 Mitigation

The optimisation of design, the strength of procedural controls, the degree of water and waste treatment facilities, and the chosen means of construction and waste disposal provide the principal means of mitigation. Procedural controls on practices such as biocide usage have a significant bearing on environmental effects in normal operation.

Site-specific fish-deterrence systems, whether acoustic, electrical or optical, are becoming commercially available, and screening technology for small-volume intakes is well established. For a direct-cooled plant on an upper estuary or river with a relatively restricted waterway, required mitigation costs will be higher than for plant in a more open situation. Plant that is not direct cooled, but uses cooling towers and lower intake volumes, will have little difficulty excluding fish from intakes, and such exclusion systems, regarded practicable by the regulators, can be assumed in most cases. Fish return or simpler trash return systems may be appropriate in some circumstances.

The need for biocides is highly site specific; application may be optimised by monitoring settlement of fouling organisms such as mussels. In some freshwater circumstances it may be necessary to dechlorinate the effluent by (say) sulphur dioxide injection.

Creative conservation, such as the establishment of new habitat types, is a valuable mitigative measure in some circumstances. Habitats that may be created include brackish water lagoons and artificial reefs. At Hinkley Point, an arrangement was made for land to be provided to compensate for an inter-tidal area lost from a site designated as being of international importance for bird life under the RAMSAR Convention on Wetlands of International Importance (1971).

7.6 SOLID WASTES

7.6.1 Existing environment

The feature of interest in this case is the physical and chemical nature of the receiving environment, especially soils and associated groundwater, into which wastes may be deposited. Characteristics of particular interest are the occurrence and leachability of trace species and their potential for bioaccumulation, persistence and toxicity.

7.6.2 Solid waste quality standards, regulations and guidelines

Much of the concern over disposal of solid wastes to land relates to the contamination of groundwater and surface water due to leaching of contaminants by rainfall or other precipitation events (NRA, 1991b).

'Controlled wastes' are dealt with under the Environmental Protection Act 1990 (previously the Control of Pollution Act 1974). Waste materials that are particularly hazardous or toxic, defined as 'special wastes' under the Control of Pollution (Special Wastes) Regulations 1980, are subject to more stringent controls than controlled wastes. Radioactive wastes are dealt with under the Radioactive Substances Act 1960, amended by the Environmental Protection Act 1990. Accumulation and disposal of radioactive wastes are covered by these Acts. The Ionising Radiations Regulations (1985) to protect the workforce include provision for the handling and storage of radioactive waste.

The concept of 'duty of care' for the management of 'controlled wastes' was instituted by the Environmental Protection (Duty of Care) Regulations (DoE, 1991b). Breach of the Duty of Care is a criminal offence. This Duty covers commercial and industrial waste transported from an originating site, with the exception of radioactive wastes and explosives.

Special wastes may also be subject to Section 34 of the Regulations, which requires all waste holders to ensure that:

- there is no unauthorised or harmful deposit, treatment or disposal of waste while in their care and/or the care of their contractors

- the escape of waste from the holder's control is prevented

- the transfer of waste is only made by an authorised person

- a written description of the waste is also transferred.

Detailed requirements relating to the preparation and transfer of documentation are given in the Regulations; Section 6 covers 'Expert help and advice'.

Demonstration of ability to comply, or undertakings to comply, with waste regulations is a part of the environmental evaluation of a project. Proposals should also take account of plans for treating and disposing of controlled wastes drawn up by the Waste Regulation Authority for the area, usually a County Council in England, a District or Borough Council in Wales and the Islands or District Council in Scotland.

An EA should include an outline of fuel transport arrangements and an assessment of the effects of such transport on the public.

Spent (irradiated) fuel is handled as radioactive material under the Radioactive Substances Act 1960, which requires registration for keeping or use of radioactive materials from HMIP (in Scotland: HM Industrial Pollution Inspectorate). Spent fuel is held at a power station in cooling ponds or dry cooling facilities, until such time as fuel element self-heating, from radionuclide decay, and fuel element radioactivity have fallen to levels at which it can be transported safely away from the site. Containers used for this operation are approved under International Atomic Energy Agency (IAEA) Series 6, Regulations for the Safe Transport of Radioactive Materials.

Accumulation and disposal of wastes also requires authorisation from the Inspectorates.

7.6.3 Impact evaluation

Electricity generation produces various controlled, special and radioactive wastes depending on the type of generation system. An EA should review the nature of waste arising and arrangements for their safe management and disposal.

In the operational phase, these wastes include:

- pulverised fuel ash (PFA)
- boiler cleaning wastes
- sludges from water treatment plant
- flue gas desulphurisation plant wastes
- fluidised bed combustion wastes
- radioactive wastes
- ash from oil-fired boilers
- polychlorinated biphenyls (PCBs – being eliminated from existing plants).

In the plant decommissioning phase the wastes may include:

- radioactive material
- asbestos
- treated timber wastes (from cooling towers)
- remaining PCBs.

Radioactive materials are produced when uranium in nuclear fuel rods is consumed in a reactor to generate heat. Solid radioactive waste materials arise from the control of radioactive discharges to air and water and in irradiated fuel storage facilities.

The most notable characteristic of solid coal-fired station waste is the very large amount of ash produced. Typical quantities for a large modern power station (2×900 MW) are given in Table 7.16. There are two basic types of ash: furnace bottom ash (FBA) and PFA. The latter is a fine powder comprising 60–80% of the total, depending on fuel quality and furnace/boiler type.

Compared with PFA, FBA is a coarse material that finds a ready market as road aggregate and in the production of thermal insulation blocks. PFA is also sold to the construction industry as fill, cement substitute and for making building blocks. However, large quantities of PFA remain for disposal or utilisation, for example to reclaim gravel and clay pits and in landscaped mounds. Another option, offshore disposal, is being phased out in the UK. Potential environmental effects also arise during the storage and transport of bulk waste material as well as during their use or disposal.

Radioactive waste is classified in three categories (Table 7.17) as follows.

- *Low-level waste* (LLW) arises mainly from operational and maintenance activities in controlled areas and comprises contaminated items such as clothing, paper towels and plastics as well as lightly contaminated oils. LLW is segregated at source into non-combustible and combustible material and is either compacted or the volume is reduced by incineration in authorised facilities or for disposal it is packaged in sealed containers and stored in shallow, concrete lined land repository at Drigg in Cumbria.

- *Intermediate-level waste* (ILW) comprises items such as fittings removed from fuel elements prior to shipping and wet wastes, including sludges, sands and ion-exchange resins from the treatment of cooling pond water or active effluents prior to discharge. ILW is also expected to arise during decommissioning of the plant. It is stored at power stations in purpose-built vaults and tanks approved by the Nuclear Installation Inspectorate. Waste may be solidified for interim storage, transport off site and disposal. UK Nirex is responsible for developing a repository for ultimate disposal of ILW and LLW, the proposed facility being hundreds of metres underground.

- *High-level waste* (HLW) comprises radioactive waste products removed from spent nuclear fuel during reprocessing (in liquified as well as solidified form) held in storage at Sellafield (Cumbria) and Dounreay (Scotland) in cooled facilities to remove heat generated by radioactive decay.

An example of estimated arisings of ILW and LLW in operation is given in Table 7.18.

Table 7.16 Estimated major raw material requirement and solid product arising during the operation of a 2 × 900 MW power station[1]

Raw materials	Average quantities (tonnes/year[1])
Coal	5,000,000
Limestone	300,000

Solid products of operation	Average quantities (tonnes/year[1])
Pulverised fuel ash	800,000
Furnace bottom ash	200,000
Gypsum	500,000
FGD waste water treatment Plant sludge arisings	10–25,000

Water requirements	Average volumes (m³/day)
Boiler feed water (potable)	9,000
Domestic water (potable)	300
FGD process water (potable)	Up to 13,000

Note:

1. Based on a 76% load factor and a coal containing 20% ash and 1.7% sulphur.

Table 7.17 Classification of radioactive waste

Waste classification	Level of hazard handling requirements	Examples
High-level waste (HLW)	Highly radioactive and significantly heat generating. Requires shielding and containment	Radioactive waste products removed from spent nuclear fuel during reprocessing (in liquid or solidified form)
Intermediate-level waste (ILW)	Less radioactive. Requires some shielding and containment	Filters at power stations and reprocessing plant. Old radiation sources from hospital equipment
Low-level waste (LLW)*	Ranges from very slight radioactive contamination to ILW level. Most requires no shielding	Waste paper and protective clothing from nuclear power stations, hospitals and industry[1]

Note:

1. Some naturally occurring materials have levels of radioactivity comparable with LLW.

Table 7.18 Estimated arisings of operational waste at an AGR reactor

Waste type	Approximate specific activity of new waste arisings (TBq/m³)	Accumulation at 1.1.86 (raw)[1]	Arisings to end of operating life (2002)[2] (raw)[1]
ILW			
Wet wastes			
- ion exchange resins	2	3	12
- sludge	30	7	19
Fuel element debris			
- stringer debris	2000	210	550
Misc. dry solids			
- misc. contaminated items	<0.4	0	51
- misc. activated components	2000	11	110
TOTAL ILW	-	230	750
LLW			
Dessicant and catalysts[3]	4×10^{-5}	30	74
Misc. LLW[4]	0.0001	13	1400
TOTAL LLW	-	43	1500

Notes:

1. Values are given as volumes in units of m³. All values, including totals are rounded to two significant figures.
2. An operating lifetime of 25 years has been assumed.
3. Current estimates indicate that desiccant and catalysts will be in the LLW category. However, depending on operational experience it is possible that they may in fact come into the ILW category.
4. Volumes given for misc. LLW are those after initial on-site processing.

There are three basic types of conventional waste treatment plant at power stations, all of which result in either used ion exchange resins or sludge residues:

- make-up plant (to purify water extracted from rivers, etc.)

- condensate polishing plant (at plant with once-through boilers needing a continuous flow of higher standard feedwater)

- flue gas desulphurisation process liquid waste treatment systems, e.g. for gypsum-producing plants (see Table 7.19).

Each of these systems where they give rise to liquid effluents is subject to the controls described in Section 7.5.2.

Table 7.19 Composition of sludge from typical FGD effluent treatment

Contaminant	Concentration (mg/kg)	Contaminant	Concentration (mg/kg)
Arsenic	432	Antimony	69
Cadmium	94	Boron	0
Chromium	874	Fluoride	51522
Copper	1148	Iron	23692
Lead	1349	Manganese	1635
Mercury	94	Molybdenum	842
Nickel	203	Selenium	280
Zinc	1811	Tin	39
Aluminium	51007	Vanadium	294

Source: Taylor *et al.*, 1988

The introduction of seawater washing systems for FGD would both add contaminants to the cooling water effluent and to a large degree sterilise the entrained flow.

7.6.4 Mitigation

Coal stocks are treated by rolling and dampening to avoid fugitive dust. To make PFA more controllable, it is conditioned prior to transport by adding 10−20% water (by weight) to improve handling and prevent wind-blown dust nuisance. Sometimes it is transferred to settling lagoons, permanently or temporarily, in which case it can be transported by pipeline after mixing with about 60% water. The provision of settlement lagoons, often requiring land take (sometimes in inter-tidal areas), results in habitat loss and may affect coastal physiography.

Ash may need to be covered with soil depending on the proposed use of the reclaimed land because of trace metal content and leachability. With care, land can be restored for agricultural or recreational use. For example, some 85 acres of worked-out pits around Didcot Power Station have been reclaimed and now produce arable crops. Use of newly reclaimed land for domestic housing development has not been recommended owing to difficulties in exercising proper control.

Another disposal method is mounding, where the land behind the working (disposal) face of a large mound is grassed over. One such mound is now grazed by sheep on the reclaimed area while disposal continues on the working face.

Where top soil is in short supply, special planting and cropping cycles have been developed to provide limited agricultural use with careful management. Care is needed because of the initial uptake, or runoff, of trace elements, including boron, chromium and selenium.

Where ash is mounded and subject to long-term exposure, a thin layer of bitumen, sometimes including grass seed, can be sprayed onto the surface to bind it and prevent wind erosion. Similarly, the introduction of vegetation to lagoon settling systems can mitigate resuspension of hollow particles (cenospheres) floating on the surface. Water extracted from lagoons to dry them out and runoff from ash traps are controlled as necessary by peripheral drains and settling tanks to remove any remaining suspended solids. The effluent is subject to Consent to Discharge.

Mitigation measures relevant to waste radioactive materials include:

- segregation in shielded facilities
- allowing time for radioactivity and heat to decay
- designation of special contamination and radiation control areas
- reduction of waste volumes
- solidification of wet wastes.

Monitoring of waste radioactivity is carried out in ways agreed with regulators and results are reported. Nuclear sites are subject to independent inspection.

7.7 NOISE AND VIBRATION

7.7.1 General and noise background

Noise is defined as 'sound which is unwanted by the recipient'. Under the Control of Pollution Act 1974, and Part 3 of the Environmental Protection Act 1990, a developer is required to employ the best practicable means for preventing, or counteracting, the effects of noise from proposed installations. An aim is to ensure that operation does not give rise to justifiable complaint from the public. The local environmental health officer should be consulted on the environmental noise protection criteria proposed for a given development.

Criteria for assessing the likely significance of an additional source of noise are based on its comparison with pre-existing background noise in the neighbourhood. Noise background is generated by natural means, such as by wind, animals, birds and by human beings, including use of machinery. It varies through the course of the day and with seasonal activities. Quietest periods normally occur at night.

For new electricity generation plant it has been common for background noise to be surveyed around the location of proposed new projects. Measurements are taken over selected 24-hour periods in conditions of low wind speed, less than 3 m/s, to ensure that noise recorded is not due to aerodynamic noise in the microphone. Noise measurement locations are selected to establish background noise levels:

- outside properties closest to the proposed power station

- in communities closest to the proposed power station

- where an operational station already exists; outside properties in areas typical of the neighbourhood but remote from the influence of the existing power station noise.

Measurements are taken over several 24-hour periods, spread out if possible to establish seasonal variability. Analysis is undertaken of measurements of tape recordings of noise levels measured during the day, evening and night-time periods. Statistical measures of noise, such as L_{95}, the background sound level in dB(A) exceeded 95% of the time, can then be established for each measurement period at each location.

7.7.2 Noise and vibration regulations and standards

A variety of regulations and guidelines deal with noise control, assessment and monitoring. Those particularly pertinent to power station projects are:

- Regulations based on EC Directive 86/188 to limit worker noise, which promotes noise control at source.

- Procedures for the description and measurement of environmental noise as contained in BS 7445 (parts 1, 2 and 3), 1991 (ISO1996 1982–1987).

- BS 5228, 1975, 1984 and 1986 (Parts 1–4), dealing with the assessment and prediction of noise radiated from construction sites. Radiation is determined from tables of noise from specific plant items. No suggestions for criteria are provided.

- BS 4142, 1990, on the means of rating industrial noise for complaint in mixed residential and industrial areas. This document is subject to continuous recent revision.

- The Control of Pollution Act 1974, and the Environmental Protection Act 1990, which provide the legal framework for the control of noise.

- A source of authoritative opinion and technical information remains the much respected Wilson Committee Report (Noise Final Report), 1963, reprinted 1973.

- World Health Organisation (WHO, 1980), which gives general guidance including sleep disturbance.

- DoE (1975) *Calculation of Road Traffic Noise* discusses traffic noise criteria and its prediction. This provides the basis for the provision of noise insulation to houses under the 1976 Noise Insulation Regulations.

- Many British Standards that need to be referred to for the measurement of noise and the suitability of instruments. Standards for measuring equipment are covered in BS 5969, 1981, 1983. A key standard is BS 4197 for precision sound level meters. There are few standards for the various types of instruments employed but this standard provides the performance specification for the important microphone and analogous amplifier sections of most instruments.

Large industries such as the electricity supply industry have used the above standards and regulations as the basis for their own procedures, introducing special methods to meet their needs based on considerable experience. For example, it is recognised that BS 4142 provides insufficient information on the measurement of background noise for formulation of complaint criteria for large installations. As noted, whichever criteria are adopted, they will be subject to agreement with local authority officers.

7.7.3 Noise impacts

Noise from site preparation and construction activities is subject to special consideration because of its temporary and variable nature. The mobility of the plant employed accounts for the difficulty in providing effective noise control.

Construction noise

Guidelines to what may be considered practicable and acceptable construction noise criteria are as follows:

- During the day, noise from a site should not exceed a level of 70 dB(A) outside nearest dwellings.

- Construction work at night should not disturb the sleep of occupants at the nearest dwellings.

This typically requires noise levels near dwellings of less than 45 dB(A). It is often practicable to limit more noisy construction work to the daytime period.

Noise emission from a site may be assessed using models for phases of the construction programme together with calculated noise emissions using BS5228 methods and theoretical and measured noise propagation from site to nearby communities under different wind conditions.

An example of conditions agreed with local authorities for a particular site is shown in Box 7.5.

Box 7.5 Example of specific noise limits

Noise emitted by on-site construction activities and equipment when measured at nearby residences under neutral weather conditions shall not exceed the following sound levels (in dBA):

	Time of day	LA_{10}	LA_{01}*
Day	0700–1900 hours	60	70
Evening	1900–2200 hours	55	60
Night	2200–0700 hours	40	45

* The LA_{01} limit serves to control impact and impulsive noise.

Operational noise

The impact of operational noise depends largely on the amount by which the level of installation noise exceeds the neighbourhood background noise.

Background noise can be characterised in various ways.

Experience in dealing with large power stations for the ESI has shown that background noise is reliably characterised by the 'A' weighted noise level exceeded for 95% of the time, i.e. LA_{95}. If at a point of interest the LA_{10} noise level from the installation is no more than the LA_{95} measured during the quietest period of the day (2400–0200 hours) plus 5 dB, complaints would not be expected, i.e. source LA_{10} does not exceed LA_{95} plus 5 dB.

This criterion, which predates BS 4142, has proved successful in application to new electricity-generating installations and in remedial noise control situations.

To ensure that agreement can be reached with the local authorities on the criterion and that it is effective in preventing complaint in all circumstances of operation of the plant, comprehensive surveys of the background noise in the area are undertaken and propagation from the installation is determined under all wind conditions. In areas of high background noise the advice given in DoE Circular 10/73 (WO Circular 16/73) relating to creeping background noise will need to be considered. DoE Circular 10/73, extended by DoE/WO Circular 1/85 (DoE, 1985), is under review as DoE/WO *Planning Policy Guidance – Planning and Noise*, Consultation Draft 1992. BS 4142 stresses that noise assessment is a skilled operation and should only be undertaken by persons who are competent in the procedures.

Tonal impulsive noise is more likely to evoke complaint than continuous broadband noise and would be subject to a significant positive weighting to the noise level when assessed for complaint. Complications and uncertainties are avoided and costs minimised by the use of suitable specifications that ensure tonal noise of plant is reduced at source.

The reduction of noise as it propagates through the atmosphere, through and around barriers, varies with frequency and each source has a unique noise spectrum. If a gross overestimation of the cumulative noise at distance where the criterion applies is to be avoided, estimation of the propagation and cumulative noise needs to be undertaken using octave bands of noise.

Because the environmental noise criterion applies to the quietest periods of the day and is very near the background noise level, it is not practical to confirm plant specifications by noise measurements at the location where the criterion applies. The auditing of plant and plant groups is undertaken nearer the installation, e.g. at 400 m from the boundary, to confirm specifications.

Most plant sources produce mid- and high-frequency noise not dissimilar to the background noise and may be rated for complaint by comparison of the 'A' weighted noise level. However, some plant such as gas turbine exhausts produce high levels of low-frequency noise not normally found in the background and it is then necessary to specify their noise in octave bands considerably reduced in the low frequencies if complaints are to be avoided.

Some sources such as boiler safety valves operate very infrequently. Human complaint reaction to some extent depends on the energy-related or time-related level of noise. Consequently, the noise of such sources may be of a higher level than that of continuous sources. For example, where the noise limit for a continuous source may be 35 dB(A), a noise limit for safety valves may be 55 dB(A) at 800 m.

Meeting an environmental noise criterion requires a consistent plant procurement specification that not only provides noise limits for plant and building structures but indicates specific techniques for noise control and the way in which they will be audited.

Wind turbine noise

Noise is the key factor in the placement of wind turbine farms and in the utilisation of the selected site. The criterion for the acceptability of wind turbine generator (WTG) noise has a different basis from that for plant that operates in no- or low-wind conditions. The background noise against which noise from the WTG is judged will normally be due to the wind in trees etc. around habitations. The way in which background wind noise and wind turbine noise at a location some distance from a turbine might vary with wind speed is indicated in Figure 7.4.

The major noise from a WTG is of course aerodynamically produced noise from the rotor blades, which is very similar in character to the background noise. Being wind dependent, both will be variable in the short and long terms. Human response to these noises will depend on the similarity between them, in their nature and noise level. Experience in this area is so far limited in the UK. One approach suggested is that the descriptors for both wind background noise and WTG noise should be the same, i.e. LA_{50}, and, from an audibility, annoyance or complaint standpoint, a suitable criterion is that:

> the LA_{50} level of the WTG should not exceed the LA_{50} of the wind background noise by more than 5 dB (DTI/BWEA, 1992).

A more general criterion, that noise from a wind farm should not exceed 45 dB(A) in the vicinity of properties in open countryside, has been adopted in Denmark (Taylor and Rand, 1991).

It must be stressed that noise siting criteria have not been widely tested in the UK. Criteria based on other established measures, e.g. L_{90} and L_{eq}, have been applied by some developers. The ultimate aim is to confirm a criterion that can be used to set separation distances between dwellings and wind turbines that are large enough to avoid complaint, but not unnecessarily large so that exploitation of the UK wind energy resource is restricted without cause.

The audibility factor of the noise is particularly important, and the noise should be strictly controlled to limit tones that would identify the noise.

Noise from current WTGs is generated by the rotor blades and the mechanical drive train. In general, aerodynamic noise from the rotor cannot be further reduced once a 'clean' aerodynamic design is evolved and the size and tip speed sensibly chosen to minimise noise at rated conditions of load. Noise from the drive train, usually containing tones, can however be controlled by careful engineering of conventional noise control measures.

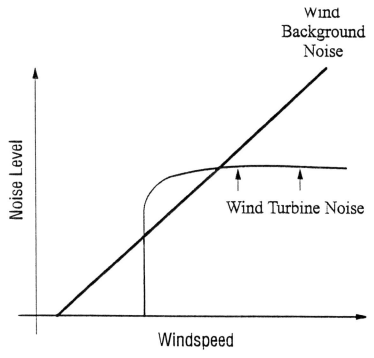

Figure 7.4 *Wind turbine and wind-generated background noise versus windspeed (turbine design and location specific)*

Auditing the noise radiation for compliance with criteria referred to habitations cannot be sensibly undertaken at these locations because of the similar level of the wind background noise. Both criteria and specifications therefore need to be referred and audited near the WTG (e.g. 1½ rotor diameters' distance) and agreed calculations for propagation employed to determine noise levels at habitations. Noise measurement in winds of greater than 3 m/s raises special problems because of wind noise in microphones. This has been discussed by the International Energy Agency (Ljunggren and Gustafsson, 1988).

7.7.4 Vibration

Generating plant operates continuously at very high loading and high speed, and vibration is minimised and continuously monitored to maintain the integrity and prolong the life of the plant. This ensures that groundborne vibration is normally environmentally insignificant.

7.8 VISUAL IMPACTS

7.8.1 General

Visual intrusion is a subjective matter, dependent on the attitude of the viewer and the nature of the surrounding landscape as well as the bulk and appearance of the power generation system.

7.8.2 Impact evaluation

A technique employed with power stations to assess potential visual impact has been to establish zones of visual influence (ZVI), i.e. areas from which the buildings could be seen, allowing for intervening obstructions.

The advent of computerised mapping assists this process. ZVIs can help in defining a set of representative locations from which to develop photomontages of the development.

Although wind turbines are relatively small, windfarms have been the subject of special interest with regard to visual intrusion, partly because the best windy sites are likely to be in upland scenic areas. Wind turbines also attract attention because of the rotation of their blades, which makes them noticeable in the peripheral vision field.

7.8.3 Mitigation

Screening by planting trees or bushes is a standard technique for reducing visual impact. Even for large structures, boundary screening can hide low-level visual clutter, such as pipes, huts and vehicles, allowing the building architecture to be appreciated with less distraction.

It is usually an architectural decision as to whether a structure should be made to stand out in the landscape, or whether cladding and colour should be used to soften visual intrusion. Blocks of a structure can be broken up visually by using different colour and texture treatments for its different parts.

For particularly sensitive locations, the view can be blocked by planting shrubs or trees close to the viewpoint. Appropriate planting and management can also add ecological interest, but permission for off-site planting depends on cooperation from landowners or local authorities, e.g. for road verges.

The Countryside Commission recommends that the installation of wind turbines should be limited to land of relatively high wind speed but outside designated areas such as Heritage Coasts and SSSIs (Countryside Commission, 1991). The size and number of wind turbines and the array layout pattern can be adjusted to suit the form of the landscape, e.g. flat coastal or rolling hills. Windfarms made up of a single type of turbine have been preferred to those based on a mixture of designs, i.e. where the number of blades per turbine or their axis and direction of rotation vary from unit to unit.

Transmission lines can also be distinctive features in the landscape. The main mitigation method is careful consideration of alternative routes. Undergrounding of lines is possible but the additional costs involved, including those for cooling, are very high compared with tower-borne conductors, though undergrounding is adopted when costs are justified by the sensitivity of the location, e.g. a heritage feature, and the degree of intrusion.

7.9 INTERFERENCE WITH RADIO COMMUNICATIONS

Parts of built structures can disturb electromagnetic communications according to the wavelength of incident signals. Examples are as follows.

7.9.1 Wind turbines

Scatter of television signals by rotating blades creates time-dependent multipath signals at a receiver. These appear as intensity-modulated ghost images, making them especially noticeable. The extent to which they prove troublesome will depend on the existing quality of reception in the area. Interference annoyance criteria are reviewed by the Comité Consultative Internationale des Radio Communications (CCIR).

Television interference can be mitigated by various means, such as re-orienting receivers where an alternative signal is available, installation of a repeater station, or resort to supply by cable. Advice is available from the BBC and IBA.

Periods of rotation of wind turbine blades tend to be comparable to time constants for gain control circuits in radio systems and may cause interference. Systems vulnerable in this respect include VHF navigational systems on aircraft.

It has been established that wind turbines should not be located close to the line of sight between microwave communication links and that siting within several kilometres of radio transmitting stations needs extra care.

A priority for scoping or assessment is to identify which radio bands in the area of the proposed windfarm are potentially at risk. The Radio Communications Agency of the DTI have information on radio services in different areas. Other organisations that may need to be consulted include British Telecom, Mercury Communications, Ministry of Defence, the Home

Office and the Civil Aviation Authority. Field surveys by specialists will help to establish the amount of electromagnetic communications traffic locally.

7.9.2 Transmission lines

These also have some potential for radio interference, the consequences of which can be mitigated by adhering to BS 5602, 1978, Code of Practice for Abatement of Radio Interference for Overhead Power Lines; this code considers the field strengths associated with corona discharge. Shielding of radio receivers due to the presence of overhead line conductors can usually be dealt with by repositioning the receiver aerial, as can reflections from tower structures. As with wind turbines, it is recommended that new transmission towers should be sited away from direct sight lines between microwave relay towers.

The National Grid Company will advise or assist if a transmission line is thought to be causing interference.

7.10 ELECTROMAGNETIC FIELDS

7.10.1 General

There is a natural magnetic field in Britain of about 40 amperes/metre (A/m) and an electric field of about 100 volts/metre (V/m) which may rise in thunderstorms to thousands of V/m.

7.10.2 Regulation and standards

Suspicions were raised in the 1960s that living in close proximity to high-voltage transmission lines or exposure to electromagnetic fields in the workplace might carry an increased risk of cancer. Following extensive research worldwide, a number of authorities, including the World Health Organisation and NRPB, have concluded that there is no established association between electromagnetic fields and cancer and do not advocate any special precautions for people living close to transmission lines. Research into health effects is ongoing in several countries by means of epidemiological studies, allowing as far as possible for alternative causes of disease.

NRPB has set guidelines on exposure from high electric and magnetic fields, based on known effects such as perceptive induced currents. These are 12,000 V/m and 1600 A/m respectively, which are greater than the highest field strengths experienced from new power lines.

Planning consent for transmission lines is controlled by local authorities subject to statutory electrical safety clearances.

7.10.3 Impact evaluation

Alternating electric current produces an electromagnetic field of some $2-11$ kV/m in a few places under the highest voltage lines (400 kV). Typically the magnetic field has a maximum value of 30 A/m below a 400 kV line. Lower-voltage lines generally produce lower field strengths. Both fields decrease with distance from a line.

7.11 HERITAGE FEATURES (ENGLAND AND WALES)

The occurrence or proximity of features such as ancient monuments, archaeological sites and listed buildings is relevant to an EA.

National organisations hold information on these topics (see Table 7.1), including the county planning departments' sites and memorial records (or similar) maintained by county archaeologists (where they exist). In some cases a nominated university takes the lead.

The Council of British Archaeology (CBA) acts as coordinator for national, county and local societies, which support the development of archaeology.

If a proposed development impinges on a site of archaeological interest, a rescue archaeology exercise may be needed prior to development to record the feature and remove items of interest. County planning departments (or their equivalent) will inspect and advise on preservation of items unearthed during excavations. Advice can also be obtained on marine archaeology features.

A summary of the results of surveys carried out in relation to a proposed power station site is given in Box 7.6. This lists six sites of archaeological interest and 18 listed buildings. Details of habitat and plant species surveys for the same site are provided in Figures 7.5 and 7.6.

Box 7.6 Archaeological and listed building surveys (example)

Sites of archaeological interest	Listed buildings
1. Site of post medieval saltern	1. Church of All Saints
2. Ashlett Mill 17th century tide mill	2. Tombchest
3. Site of post medieval saltern	3. Tombchest
4. Calshot Castle 16th century artillery fort	4. Nos. 74, 76 Church Lane
5. Flying boat station	5. Falcon Hotel
6. Site of post medieval saltern	6. Haywards butchers
	7. The Jolly Sailor Public House
	8. Ashlett House
	9. Seaview cottages
	10. Farm range, Badminston Farm
	11. Badminston Farmhouse
	12. Ower Farmhouse
	13. Nos. 1, 2 Nelson Lodge
	14. Garden walls and Temple Eaglehurst
	15. Eaglehurst House
	16. Flight of steps to gateway, Eaglehurst
	17. Luttrells Tower, Eaglehurst
	18. Calshot Castle

7.12 ECOLOGICAL APPRAISAL: TERRESTRIAL

7.12.1 General

The area of interest for the terrestrial ecology is the proposed site, the adjacent land used during construction and a buffer zone around these areas. The buffer zone provides a basis for evaluating natural trends over the operating life of the station. Road and rail access routes should also be surveyed. Initial scoping surveys are carried out at the early planning stage if there are several potential site locations and/or scheme layouts in the locality. The information will derive from existing records, possibly supplemented by a Phase 1 (English Nature) type survey; this is a field survey of basic vegetation types. Effects on migratory and local birdlife are also of interest. The prime source of ecological data will be the appropriate regional office of English Nature (or its equivalents). Other sources of documented data are given in Table 7.20. If a Phase 1 survey has not been done previously for the area of interest or if an available survey is dated, then one should be carried out by a trained observer. This survey involves the classification and mapping of basic vegetation types such as scattered scrub, unimproved grassland, rich fen or vegetated shingle and mixed woodland. An example of general habitat mapping for an EA is given in Figure 7.5.

Table 7.20 Sources of existing ecological information

General ecological information (including both wide-ranging and site-specific data)

    English Nature (and equivalents)
    Local planning authority
    County (or Scottish) Wildlife/Nature Conservation Trust
    County Biological Records Centre
    Local and regional museums
    Local university departments

Information on species groups

    Royal Society for the Protection of Birds
    County bat or badger groups (may be attached to County Wildlife Trust)

Information on particular habitats

    British Trust for Ornithology (estuary birds)
    Wildfowl and Wetlands Trust (wetlands)
    National Rivers Authority (rivers)
    MAFF

Note:

This list is not comprehensive. Often these contacts will lead to others, perhaps including local amateur naturalists, with valuable information. See also Table 2.2.   Source: Henderson, A., 1990,

A Natural Vegetation Classification (NVC) has been developed (Rodwell, 1991, 1992a, 1992b), extending broad classifications set by the Nature Conservancy Council (NCC, 1990). Dominant plant species and notable localised features are recorded in surveys, most easily in summertime. Of prime interest are specially designated areas in the vicinity of the site where there are various sorts. These fall into three main groups (Table 7.21).

Phase 2 surveys are more detailed and time consuming and should be carried out by trained staff. Several specialists may be involved at any one site bearing in mind the key habitat types relevant to that project (e.g. lowland heathland, sand dunes or saltmarsh).

Species fall into standard classes, such as vascular plants, lower plants, invertebrates and birds. Some species are particularly sensitive to disturbance or to pollution, such as lichen. Survey techniques vary with the group of interest. Bird surveys are carried out in the breeding season to establish territories, and migratory birds are recorded in winter. Time has to be allowed for such surveys to be done unless existing records are adequate. Ideally, the surveys would be spread over a complete year, starting around the beginning of a season.

7.12.2 Impact evaluation

Evaluation of the significance of the disturbance of different habitat-types can be a challenging exercise. Nature conservation values of communities of species are evaluated on the basis of nine main attributes, namely:

- size (extent)
- diversity (species and communities)
- rarity (species and communities)
- fragility
- typicalness
- recorded history
- position in an ecological/geographical unit
- potential value
- intrinsic appeal.

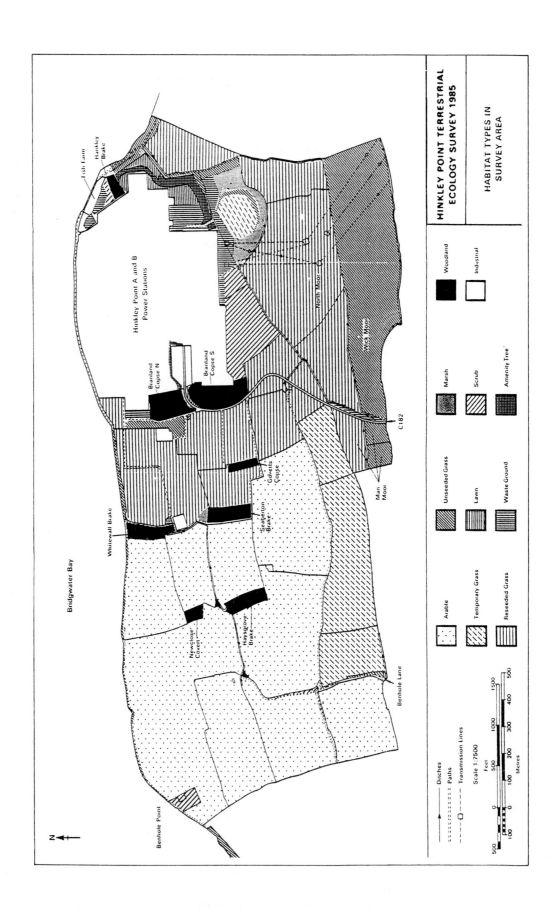

Figure 7.5 *Habitat types in survey area*

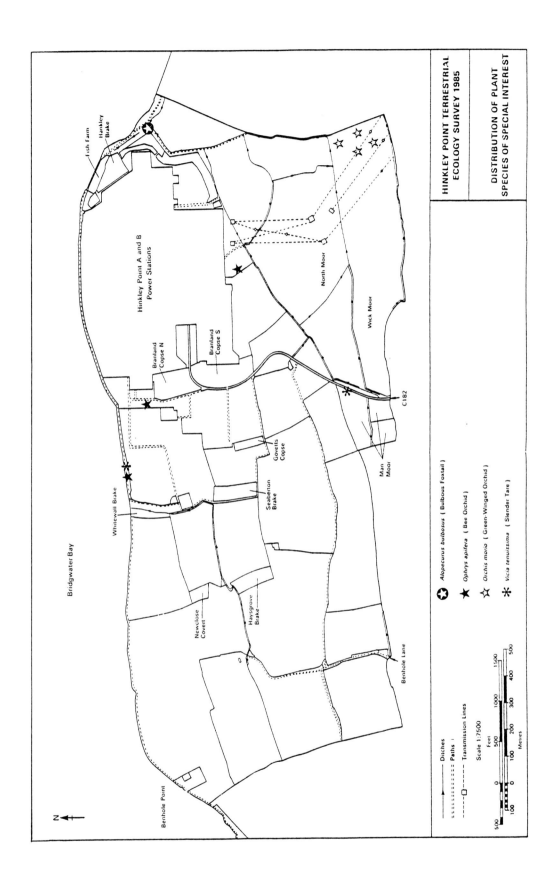

Figure 7.6 *Plant species in survey area*

Table 7.21 Nature conservation designation: main types

1. Statutory sites notified under the Wildlife and Countryside or previous Acts. These include SSSIs, statutory Local Nature Reserves and National Nature Reserves. Some of these may also be Ramsar sites (The Convention on Wetlands of International Importance especially as Waterfowl Habitat), or Special Protection Areas (EC Directive 79/409 on the Conservation of Wild Birds) or Special Areas for Conservation (Directive 93/43).

2. Non-statutory sites designated by voluntary nature conservation bodies. These could include nature reserves managed for conservation, and other land that is of interest but is not notified as SSSI.

3. Areas and sites allocated for protection by the local planning authority, in Structure and Local Plans. These may include both broad swathes of land and small defined areas. Plans often have policies relating to SSSIs and to non-statutory sites.

Source: Henderson, A., 1990.

Quantifiable factors are size, diversity and rarity, whereas type and position in an ecological/geographical unit are assessed in relation to the NVC. The other attributes have to be based on value judgements.

An example of various sorts of nature conservation interest is shown in Table 7.22, in this case for an estuarine site. This sort of information is used variously to adjust the layout of the project or the area disturbed by construction activities, to identify habitat types that might be reproduced elsewhere to compensate for losses and to identify protected species that can be moved elsewhere (e.g. Great Crested Newts). Such exercises have to be carried out carefully and, for notified species, only with the involvement and agreement of English Nature or its equivalents.

In some cases, rarity is assessed in terms of the number of 10 km squares in the UK in which the species has been recorded, e.g. for vascular plants. The *Atlas of Breeding Birds in Great Britain* is also a useful source of information (BTO, 1993). In some cases, ancient woodland has been mapped on a county scale. Hence the information available for evaluation may differ from site to site. In order to judge nature conservation value, it may be helpful to consider 'value' in terms of local, regional, national or international interest. An example of a mapping exercise for plant species of special interest is presented in Figure 7.6.

International interest for bird life arises where a significant proportion of the international community of a migratory bird species regularly visit a specific site (e.g. shelduck in Bridgwater Bay). The importance may be acknowledged by designation of the area as a site of international importance under the Ramsar Convention on the Conservation of Wetlands. Various Directives and Conventions relate to wildlife protection, including the EC Birds Directive with its provisions for special protection areas and the Berne Convention on Threatened Species and Habitats. Electricity undertakings have responsibilities for conservation under Schedule 9 of the Electricity Act 1989.

Table 7.22 Examples of major habitat types (case study)

Habitat types	Nature conservation interest
Arable	Various crop species, including wheat, barley, maize and temporary ley. Relatively low wildlife interest
Grass: Permanent improved	Species-poor grasslands used for grazing or silage. Wildlife interest generally low but dependent upon management regime and associated habitat types
Grass: Permanent unimproved	Floristically and structurally diverse due to edaphic factors, longevity and management. One meadow is a County Heritage Site. Relatively uncommon habitat-type of locally high (possibly regional) nature conservation value
Amenity grasslands	Road side verges, lawns, public open spaces. Limited wildlife interest due to management regime usage
Woodland	Primarily semi-natural mixed deciduous woodlands of various ages, extent and species diversity, plus small areas of plantation and amenity planting. Not uncommon with local area. Typical lowland woodland breeding bird community. Locally medium conservation value
Wetland, carr and scrub	High structural and species diversity related to edaphic factors, adds to overall diversity of habitat types in the area. Local medium conservation value
Reclaimed wetland	Mosaics of reed beds, pools, grasslands and scrub invasion by sallow, bramble and birch. Typical wetland breeding bird community. Uncommon paramaritime plant species present. Areas of relatively high nature conservation value
Hedgerows	Managed field boundary hedgerows. Primarily with standard oak trees offering habitats for breeding birds; act as corridors between habitat-types. Locally medium value
Saltmarsh/ intertidal area	Part of SSSI. Diminishing regional resource due to natural erosion, reclamation and invasion of *Enteromorpha*. Roosting area for overwintering waders and wildfowl. Inter-tidal area invertebrates represent main food source for over-wintering bird species. Regionally important area for waterfowl: seven species present in nationally important numbers, one present in internationally important numbers. Regionally high wildlife value
Disturbed ground	Structurally diverse due to disturbance. Characteristic colonising flora. Areas of scrub, good breeding bird habitat, and grasslands, attractive to butterfly species. Locally medium value
Tracks	Roadside verge flora with hawthorn, oak, ash, elm hedgerows. Little botanical interest
Shingle	Characteristic flora with two *Red Data Book* species. Uncommon habitat type of regional value

7.12.3 Mitigation

Various methods can be employed to mitigate the effects of a development. An example is given in Table 7.23.

One method is 'creative conservation' (NCC, 1984), which seeks 'to enlarge the resources of nature by re-creating habitats and communities or re-introducing species lost through human impact'. The likelihood of success of some conservation projects may need to be demonstrated by pilot studies. Creative conservation is not a universal panacea; some habitat types such as ancient woodlands are irreplaceable. Nevertheless, power stations can in some circumstances, like other industrial developments, provide havens for wildlife. For example, the power station site at Dungeness is the national breeding site for the Sussex Emerald Moth.

Table 7.23 Examples of impacts of power stations on ecology

Habitat type	Aspect of development	Effect	Mitigation
Dune grassland and vegetated shingle	Beach and foreshore works during construction	Temporary loss of disturbed habitat with conservation interest	Restoration methods to be applied
Reed bed	Access and temporary haul road	Disturbance to small part of area	Bridging where practicable. Maintenance of normal water flows during construction and operation. Removal of temporary structurs to allow recolonisation. Planting where necessary
Conifer plantation	Access and temporary haul road	Permanent loss of low conservation interest habitat type	Create road margin interest and diversify the species and age structure of the adjacent plantations to produce habitats of higher conservation interest
Mixed plantation	Access road	Loss of small area of limited interest	Create road margin conservation interest
Arable	Access road and temporary contractors areas	Permanent and temporary loss of low conservation habitat	Restoration to agriculture or semi-natural vegetation such as woodland and heathland if practicable to increase conservation interest
Hedges	Temporary contractors area	At least partial loss of habitat type of local conservation interest	Retain hedges around boundaries and where practicable. Replant hedges if land returned to agriculture

Other measures deserving consideration in appropriate circumstances include:

- creative conservation projects in land disturbed by construction activities;

- markers to deter birds from transmission lines (e.g. plastic spheres)

- provision of alternative breeding sites (e.g. for owls)

- relocation of protected species (e.g. Great Crested Newts) with the agreement of English Nature or its equivalents

- timing of activities to reduce disturbance

- provision of alternative protected territory, e.g. in compensation for loss of part of a Ramsar Site at Hinkley Point.

Although various practical wildlife protection methods of this type have been tried, experience with them is largely still limited. They can be expected to evolve with time.

7.13 OTHER SOURCES

A new generation of fossil systems, natural gas, coal and refuse burning is under development. Most have environmental advantages over present systems in reduced emissions to atmosphere, including less carbon dioxide. Some, such as fluidised beds, will produce more solid waste.

Other developing technologies are the 'renewable' processes including biomass. Their potential is discussed by REAG (1992). A number of these are listed below, with their main potential impacts. In all cases it is expected that environmentally acceptable facilities are possible, though in some cases the costs of environmental controls are as yet unclear. In others, such as wind, biomass and photovoltaic systems, public reaction to their appearance on a significant scale in the landscape has yet to be tested.

Some potential environmental impacts and risks to be controlled, and environmental benefits, for a range of new developing resources are given below. These are summarised in Box 7.7.

7.13.1 Biomass crops for energy

There are risks to workers in harvesting and transport. Monoculture may encourage pathogens and pests and pesticides; herbicides and fertilisers impact on surface water and groundwater. Relatively high organic polycyclic aromatic hydrocarbon (PAH) emissions are controlled by the management of combustion, although associated sulphur dioxide emissions are generally low. There are no net carbon dioxide emissions where crop photosynthesis balances combustion. Site selection should avoid ecologically sensitive locations.

7.13.2 Photovoltaics

The thin-film materials utilised include toxic elements (selenium and tellurium). If the technologies become widespread, hazard control will be an issue for raw materials extraction, manufacture, waste handling and disposal. Gallium arsenide systems represent a greater potential hazard than silicon technology. Large area projects (sq.km.) will affect habitats and have visual impact. This can be avoided by installation of pv energy collectors on buildings.

7.13.3 Landfill gas

This has the advantage of consuming methane, some of which otherwise leaks into the atmosphere where it is a more powerful greenhouse gas than carbon dioxide. Individual installations are sized to generate a few MW.

7.13.4 Refuse-derived fuel

Refuse-derived fuel (RDF) combustion plant tends to be relatively small scale (up to 10 MW th). A number of projects are under active development. Assuming the scale of the collection and transport of the waste is not significantly greater than for normal refuse collection activities, the main potential impact is emission to the atmosphere of pollutants associated with incineration, i.e. dioxins and related aromatic compounds, heavy metals (such as Pb, Cr, Cu, Mn, Ni, As, Cd and Hg) and acid gases (HCl, H_2S, SO_2).

Control of emissions of aromatic, potentially carcinogenic compounds is exercised by ensuring adequate fuel combustion, e.g. a minimum temperature for a long enough period at a minimum oxygen concentration (residence times would typically not be less than two seconds). Emissions should comply with limits on organic compounds and carbon monoxide.

Criteria for municipal waste incinerators were laid out in EC Directive 89/369/EEC and recent UK experience is reflected in the HMIP Chief Inspector's Guidance Notes (see Section 7.4).

Box 7.7 Key environmental issues for selected electricity generation technologies

Coal	Global warming	CO_2, methane from mining
	Acid deposition	SO_2, NO_x, HCl
	Health	NO_2, ozone, particulates
	Wastes	bulk
Oil	Global warming	CO_2
	Acid deposition	SO_2, NO_x
	Health	NO_x
Gas	Global warming	CO_2, methane leakage
	Acid deposition	NO_x,
	Health	NO_x, ozone
Nuclear	Wastes	radioactive materials
	Safety	
Wind (on land)	Visual/land use	
	Noise	
	EMI	interference
Wind (offshore)	Birds(?)	strike damage
	EMI	interference
RDF	Health	toxic trace metals, organics
	Wastes	toxic trace metals, organics
Biomass	Safety	accidents
Fuel crops	Visual/land use	
	Transport	bulk
	Health	NO_x, organics
Solar (photovoltaic)	Health	toxic gases in production
	Visual/land use	(unless installed as cladding on building)
	Wastes	toxic species
Solar (focusing)	Safety	reflected light
Hydro	Safety	(dam collapse)
	Land use	
	Health	pests in reservoirs
	Fisheries	turbine damage
Tidal	Ecosystems	intertidal habitat
	Fisheries	turbine damage
	Water quality	flow patterns changed
Geothermal	Wastes	salts, sludges with toxic species
	Health	H_2S
Sea waves	Safety	shipping
	Fisheries	
General	Ecology	construction, operation
	Archaeology	construction, operation
	Noise	construction, operation
	Safety	fuel supply, construction, operation

Control via fuel or furnace additives is the subject of ongoing research. A variety of combustion emission controls are available, e.g. sorbent injection systems, which remove acid gases as well as particulates. It has been suggested that numerous techniques could satisfy the necessary abatement performance but economic viability may be compromised unless low-cost technologies become available (van Santen and Landy, 1990).

RDF can be screened to remove metals and plastics, which burn to form organic gases. Combustion ash disposal needs care as the leachate from such a waste may contain heavy metals.

7.13.5 Geothermal

The operation of geothermal plant may result in emissions of hydrogen sulphide. Sludges arising from spent hydrothermal fluids may be toxic and include mercury.

7.13.6 Hydro power

This is an existing, widespread 'renewable' technology.

Run-of-river abstraction is now favoured technically in preference to dams and reservoir storage. The hydrology of the river between intake and outfall is modified depending on the means of abstraction: the morphology of the river bed, dependent plants and migratory fish are potentially affected (Shaw, 1993). Impingement of fish in turbines is a recognised impact. Buried pipelines minimise landscape impact but may affect ecology. Turbine houses in vernacular style blend more readily into the locality, as do locally underground transmission links. Intakes and outfalls should be screened to exclude fish, vegetation and larger sediments. Fish passes/ladders permit bypass of fish but are of variable efficiency. Individual installations are typically sized to generate up to about $1-2$ MW.

7.13.7 Tidal power

Schemes from under 50 MW to over 5000 MW are under consideration in UK. The essential difference between them is the area of tidal estuary contained by the barrage. Environmental impact is influenced by the extent of change to the tide regime in the contained basin area, but seaward effects are also important. Reduced tidal flows mean lower velocities, reduced dispersion, less sediment transport, lower turbidity and less variation in salinity. The durations and frequencies of different water levels affect intertidal saltmarshes and land drainage outfalls (STPG, 1989). Some wave action will be reduced by the presence of a barrage but a reduced range of water levels in basin area will focus this over a smaller inter-tidal zone. Waves and currents determine inter-tidal profiles, sediment substrates, and turbidity, while nutrient supply influences food chain productivity (STPG, 1993). Potentially complex water quality and ecological interactions during barrage construction and operation have to be quantified, as do the effects of dredging activities and the transport and storage of construction materials. The landscape impact of a barrage and its associated works have to be resolved. Non-energy issues including public roads, associated building land use and employment are also relevant. Environmental interactions are complex. Acoustic fish deterrence systems are being investigated: these may mitigate against fish impingement damage. Opportunities exist for creative conservation (Shaw, 1994).

7.13.8 Sea wave power

Installations could be coastal, near-shore or offshore (Thorpe and Picken, 1993). Absorption of incoming wave energy diminishes the transmitted resource, with consequences for the environment of water bodies in the lee, for coastal morphology and coastal ecosystems. There will be some disturbance due to construction but operational impacts for larger systems will tend to be associated with maintenance facilities. Structures off the coastline are a potential shipping hazard. Failure of moorings and anchors of structures presents safety risks for shipping and coastal structures. The accidental release of oils and lubricants from structures and their consequences must be considered, as would the scale and location of graving docks and maintenance areas. A positive benefit, incidental to the design and common to other options such as offshore wind turbines, would be the provision of effective 'artificial reefs', both fostering local fish populations and providing a haven protected from fishing.

7.14 CONCLUSIONS

The principal environmental issues associated with specific types of power station have been considered in this chapter. Key areas for different technologies are summarised in Section 7.13. The process of preparing an assessment of any scheme has been developed from more than 20 years of design, planning, evaluation and negotiation of projects by the supply industry. Thus the more recent requirement for a formal ES can readily be accommodated and the experience translated into the process of environmental assessment.

The appraisal process is continually developing as new demands come upon it in response to evolving legislation, new technologies and progressively better scientific information and assessment methodologies. These developments can be accommodated within the basic principles outlined in this chapter, thus providing a common, high standard of environmental protection important in an industry with significant potential for inducing environmental change.

7.15 REFERENCES

BARRETT, G.W. (1979)
Methods of calculating, measuring and monitoring air pollution caused by electric power stations in England and Wales
Clean Air, Vol 9, No 4, p 119

BRITISH TRUST FOR ORNITHOLOGY (1993)
The Atlas of Breeding Birds in Great Britain
BTO, Norfolk

CHIGNELL, R. (Ed.) 1986
The IEA's expert group study on recommended practices for wind turbine testing and evaluation: 5. Electromagnetic interference
Preparatory Information Issue No. 1, IEA

COUNTRYSIDE COMMISSION (1991)
Wind Energy Development and the Landscape
Technical report, Cheltenham, Glos.

DEPARTMENT OF THE ENVIRONMENT (1975)
Calculation of Road Traffic Noise
HMSO

DEPARTMENT OF THE ENVIRONMENT (1985)
Planning and Noise
Circular 10/73 and its updates, HMSO

DEPARTMENT OF THE ENVIRONMENT (1990a)
Code of Practice on Duty of Care in the Management of Controlled Waste

DEPARTMENT OF THE ENVIRONMENT (1990b)
UK Guidance Note on the Ministerial Declaration

DEPARTMENT OF THE ENVIRONMENT (1991a)
Acid Rain – Critical and Target Load Maps for the United Kingdom
HMSO

DEPARTMENT OF THE ENVIRONMENT (1991b)
Environmental Protection (Duty of Care) Regulations
HMSO

DEPARTMENT OF THE ENVIRONMENT/WELSH OFFICE (1991a)
Environmental Assessment: A guide to the procedures
HMSO

DEPARTMENT OF THE ENVIRONMENT/WELSH OFFICE (1991b)
Integrated Pollution Control: A practical guide
HMSO

DEPARTMENT OF THE ENVIRONMENT/WELSH OFFICE (1991c)
Renewable Energy
Planning Policy Guidance Note No. 22, HMSO

DTI/BWEA WIND TURBINE NOISE WORKSHOP
IREDALE R.A. (1992)
Annoyance rating for wind turbine noise
Wind Engineering, Vol. 16, no. 2, pp 109-116

GOVERNMENT STATISTICAL SERVICE (1990)
Digest of Environmental Protection and Water Statistics
GSS 13, HMSO

HENDERSON, A. (1990)
Ecological evaluation of land surrounding industrial installations
Seminar on Risk Assessment of Major Accidents to the Environment, IBC Technical Services

HMIP (1993)
Guidelines on discharge stack heights for polluting emissions
(D1 aimed at Local Authorities), HMSO

INTER-GOVERNMENTAL PANEL ON CLIMATE CHANGE (1990)
Climate Change: The IPCC Impacts Assessment
Report of Working Group 2, Australian Government Publishing Service, Canberra

INTER-GOVERNMENTAL PANEL ON CLIMATE CHANGE (1992)
Climate Change 1992: Supplementary Report to the IPCC Scientific Assessment
WMO/UNEP, Cambridge University Press

LIM, Y. (1979)
Trace Elements from Coal Combustion – Atmospheric Emissions
IEA Coal Research

LJUNGGREN, S. and GUSTAFSSON, A. (1988)
Recommended Practices for Wind Turbine testing: 4. Acoustics Measurement of Noise Emission from Wind Turbines
IEA Programme for Research and Development on Wind Energy Conversion Systems

NATIONAL RADIOLOGICAL PROTECTION BOARD (1981)
Living with Radiation

NATIONAL RIVERS AUTHORITY (1990)
Discharge Consent and Compliance Policy: A Blueprint for the Future
(The Kinnersley Report) NRA Water Quality Series No. 1

NATIONAL RIVERS AUTHORITY (1991a)
Proposals for Statutory Water Quality Objectives
NRA Water Quality Series No. 5

NATIONAL RIVERS AUTHORITY (1991b)
Policy and Practice for the Protection of Groundwater
Draft for Consultation, NRA, November 1991

NATIONAL SOCIETY FOR CLEAN AIR (1991)
Pollution Handbook
National Society for Clean Air and Environmental Protection

NATURE CONSERVANCY COUNCIL (1984)
Nature Conservation in Great Britain
NCC (Engligh Nature) Peterborough

NATURE CONSERVANCY COUNCIL (1990)
Handbook of Phase 1 Habitat Survey
NCC (English Nature) Peterborough

RATCLIFFE, D.A. (1977)
A Nature Conservation Review
Vols 1 and 2, Cambridge University Press

REAG (1992)
Renewable Energy Advisory Group: report to the President of the Board of Trade
Energy Paper 60. Department of Trade and Industry, London, HMSO

ROBERTS, M., BULL, K.R. and HULL, J.R. (1992)
Ecological effects of the critical loads approach
Clean Air, Vol 22, No 2, p 73–85

RODWELL, J. (1991)
National Vegetation Classification
Vol 1 (Woodlands), Cambridge University Press

RODWELL, J. (1992a)
National Vegetation Classification
Vol 2 (Heaths), Cambridge University Press

RODWELL, J. (1992b)
National Vegetation Classification
Vols 3-5 (in preparation)

SEVERN TIDAL POWER GROUP (1989)
The Severn Barrage Project
(with UK Department of Energy and CEGB) HMSO Energy Paper No. 57

SEVERN TIDAL POWER GROUP (1993)
Severn Barrage Project: Further environmental and energy studies
ETSU TID4099

SHAW, T.L. (1993)
The environmental effects of hydro-power schemes
Proceedings of the Institution of Electrical Engineers, No. 1, January, pp 20–23

SHAW, T.L. (1994)
The impact of barrages on a maritime environment
Proceedings of the Institution of Civil Engineers (in press)

TAYLOR, D.A. and RAND, M. (1991)
How to plan the nuisance out of wind energy
Town and Country Planning, May, p 152–155

TAYLOR, M.R.G., HEATON, R. and BATY, R. (1988)
Impact of flue gas desulphurisation on the water environment
Proc. Conf. Institution of Water and Environmental Management

THORPE, T.W. and PICKEN, M.J. (1993)
The environmental effects of sea wave power schemes
Proceedings of the Institution of Electrical Engineers, No. 1, January, pp 63–70

UK CLIMATE CHANGE IMPACTS REVIEW GROUP (1991)
The Potential Effects of Climate Change in the UK
HMSO

UK PHOTO-CHEMICAL OXIDANTS REVIEW GROUP (1987)
Ozone in the United Kingdom
First Report, Harwell Laboratory

UK PHOTO-CHEMICAL OXIDANTS REVIEW GROUP (1990)
Oxides of Nitrogen in the United Kingdom
Second Report, Harwell Laboratory

UK STRATOSPHERIC OZONE REVIEW GROUP (1990)
Stratospheric Ozone, 1990
HMSO

UK STRATOSPHERIC OZONE REVIEW GROUP (1991)
Stratospheric Ozone, 1991
HMSO

VAN SANTEN, A. and LANDY, M.P. (1990)
The abatement of emissions from small scale waste combustion
Proc. Int. Conf. on Power Generation and the Environment, Institution of Mechanical Engineers, pp 243–251

WILSON COMMITTEE (1963)
Noise - Final Report
HMSO (reprinted 1973)

WORLD HEALTH ORGANISATION (1980)
Environmental Health Criteria: 12 – Noise
UNEP/WHO, Geneva

CIRIA Research Project 424

Environmental Assessment

A guide to the identification and mitigation of environmental issues in construction schemes

8. Minerals extraction

Richard Clough, Chief Scientist, Wimpey Environmental Limited

8 Minerals extraction

8.1 TYPE OF SCHEME CONSIDERED

This chapter deals primarily with the land-based surface extraction of solid minerals in the UK. It does not cover peat extraction, deep mining or offshore aggregate extraction.

According to a recent environmental report, Britain has more than 2500 pits and quarries (Groundwork, 1991). The minerals industry encompasses some 35 different commodities, of which the most important are coal, sandstone, igneous and metamorphic rock, limestone, sand and gravel, chalk, common clay and shale. Other minerals include fluorspar and iron ore.

Table 8.1 shows the types of surface mineral surface extraction that exist and which will form the main subject of this chapter.

Surface minerals may be won by two main techniques and the difference between the techniques is probably more significant for the environment than the difference between the type of mineral being won. The main methods of land-based surface minerals extraction are as follows.

- *Quarrying by blasting*. This type of extraction is used to win minerals in solid formations rather than in granular deposits. A working face is created using explosives. Holes are drilled several metres back from the existing face. Explosive charges are placed in these holes. When the charges are fired the mineral is displaced forward and broken up and a new working face is created. High faces are worked in a series of benches. This method can be used to quarry progressively into the side of a hill or to quarry downwards into an open pit.

- *Minerals winning by mechanical excavation*. This technique is used for sand, gravel and other granular deposits. Material is won by an excavator, face shovel, backacter, dragline etc. Because of problems of slope stability, working faces tend to be less high than where the mineral is in solid formations. Progressive restoration is common with this type of extraction.

By the nature of surface mineral extraction, quarries are only temporary activities and new sources of minerals must continually be sought. Therefore, seeking planning permission is an important part of most quarry companies' business. Dealing with planning applications for minerals extraction is again an important part of the planning duties of the relevant authorities. At present, the planning authority for surface minerals extraction will be the local minerals planning authority (MPA). These authorities come under the control of local government and are organised mainly at county level. They are staffed by people with a good technical knowledge of minerals extraction. This industry-specific expertise on the part of the planning authority makes dealing with planning applications rather different from that of general planning departments who deal with many different industries.

Because of the nature of minerals winning, environmental matters have always been very important and specialist reports on the various aspects of environmental impact have been submitted with planning applications for many years, even before the term Environmental Impact Assessment came into general use. When planning applications go to public inquiry, a large part of the inquiry is always concerned with environmental matters.

Table 8.1 Examples of types of surface mineral extraction for which planning permission will be sought

Mineral	Typical promoter	Product and principal uses
Opencast coal	British Coal Opencast Executive	Opencast coal for energy supply
Sandstone. Igneous and metamorphic rock	Private quarry company	Crushed rock as aggregate for concrete and general construction. Dimension stone. Roadstone. Rail ballast
Limestone	Private quarry company	Crushed rock as aggregate for concrete and general construction and cement production. Dimension stone. Roadstone. Rail ballast
Chalk	Private quarry company. Cement company	Cement manufacture
Clay and shale	Private quarry company. Cement company. Brick company	Brick production, other clay goods. Cement production
Sand and gravel	Private quarry company. Concrete company	Aggregate for concrete, asphalt and general construction
Silica sand	Private quarry company	Glassmaking. Foundries
Metalliferous minerals	Private quarry company	Foundries. Metal production

8.2 KEY STANDARDS, LEGISLATION AND SUPPORTING ADVICE

8.2.1 Environmental Assessment requirements

The environmental impacts of minerals extractions come under the Town And Country Planning (Assessment of Environmental Effects) Regulations 1988 (SI No. 1199). Further details are given in Chapter 1. Minerals extraction will come under Schedule 2 to the Regulations. A development listed in Schedule 2 will only require an environmental assessment if it is likely to have a 'significant effect' on the environment. DoE circular 15/88 gives indicative criteria for identifying Schedule 2 projects requiring an environmental assessment. The general criteria are given in Chapter 1. The criteria for the extractive industry are given in Box 8.1.

The Department of Environment booklet giving guidance on the environmental assessment procedures contains useful flowcharts outlining the procedures to be followed. Figures 8.1 and 8.2 give flowcharts based on the ideas in the DoE publication but with procedures specific to minerals extraction.

8.2.2 Legislation affecting minerals extraction

Because of the importance of planning applications and environmental control in the minerals industry, a number of specific planning guidance policy notes have been issued by the DoE. These are known as Minerals Planning Guidance Notes and are listed in Box 8.2. There are also a number of general Planning Guidance Policy Notes issued by the DoE and the key ones of interest to minerals planning are also included in the box.

There are also two MPGs at draft stage (September 1992):

- Draft MPG 11: Control of noise at surface mineral workings. This is the first MPG to address a single specific environmental impact. It gives fairly detailed advice on how to predict, monitor and minimise noise.

- Draft MPG X: The provision of silica sand in England and Wales. This is at the limited circulation stage. It will be issued for public comment early in 1993.

Box 8.1 Criteria for Environmental Assessments in the minerals industry

> Whether or not mineral workings would have significant environmental effects so as to require EA will depend upon such factors as the sensitivity of the location, size, working methods, the proposals for disposing of waste, the nature and extent of processing and ancillary operations and arrangements for transporting minerals away from the site. The duration of the proposed workings is also a factor to be taken into account.
>
> It is established minerals planning policy that minerals applications in national parks and areas of outstanding natural beauty should be subject to the most rigorous examination, and this would normally include EA.
>
> All new *deep mines*, apart from small mines, may merit EA. For *opencast coal mines* and *sand and gravel workings*, sites of more than 50 ha may well require EA and significantly smaller sites could require EA if they are in a sensitive area or if subjected to particular obtrusive operations.
>
> Whether *rock quarries* or *clay operations* or other mineral workings require EA will depend on the location and the scale and type of the activities proposed.
>
> For *oil and gas extraction* the main considerations will be the volume of oil or gas to be produced, the arrangements for transporting it from the site and the sensitivity of the area affected. Where production is expected to be substantial (300 tonnes or more per day) or the site concerned is sensitive to disturbance from normal operations, EA may be necessary. *Exploratory deep drilling* would not normally require EA unless the site is in a sensitive location or is unusually sensitive to limited disturbance occurring over the short period involved. It would not be appropriate to require EA for exploratory activity simply because it might eventually lead to production of oil or gas.

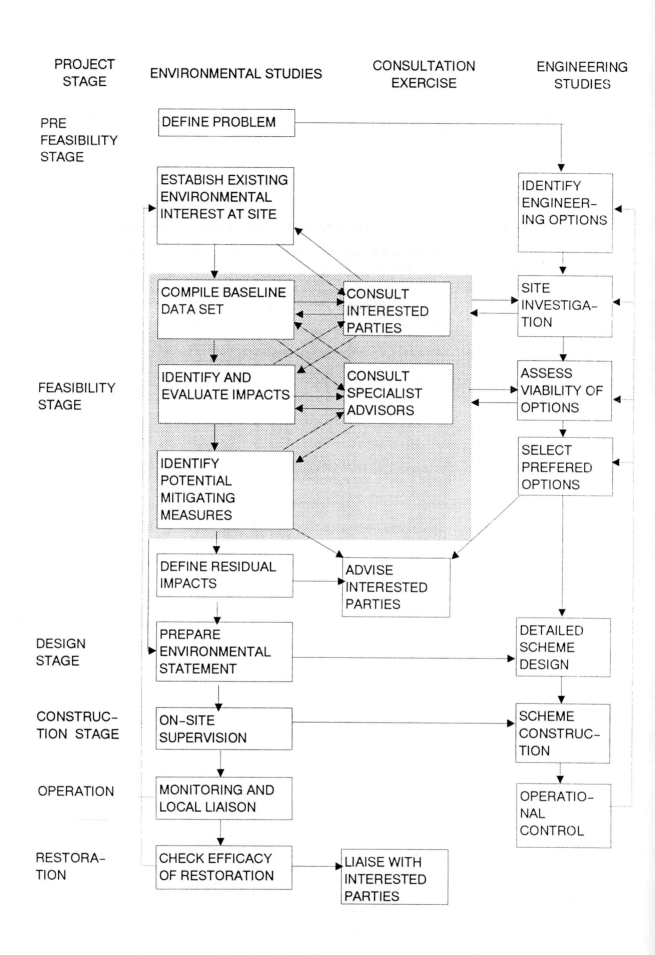

Figure 8.1 *The environmental assessment process (after Brooke and Whittle, 1990)*

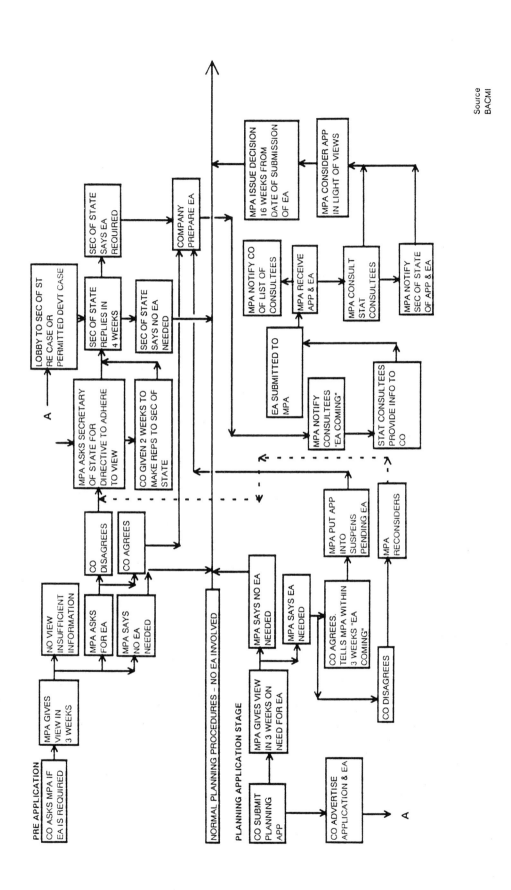

Figure 8.2 *Minerals planning applications*

Box 8.2 Official publications on minerals Environmental Assessment

Department of the Environment (1988) Circular 15/88:
Environmental Assessment, HMSO, London.

Department of the Environment (1988) Minerals Planning Guidance 1:
General Considerations and the Development Plan, HMSO, London.

Department of the Environment (1988) Minerals Planning Guidance 2:
Applications Permissions and Conditions, HMSO, London.
This gives advice on the imposition of planning conditions including conditions to control environmental impact. Many of the provisions are written in fairly general terms and more detailed specific advice is given in other MPGs.

Department of the Environment (1988) Minerals Planning Guidance 3:
Opencast Coal Mining, HMSO, London.
This states the national policy considerations for opencast coal planning and also gives a recommended approach to be taken by MPAs in determining planning applications for opencast developments.

Department of the Environment (1988) Minerals Planning Guidance 5:
Minerals Planning and the GDO, HMSO, London.
This document covers aspects of the General Development Order which are of special relevance to mineral interests.

Department of the Environment (1988) Minerals Planning Guidance 6:
Guidelines on Aggregate Provision in England and Wales, HMSO, London.
This document advises minerals planning authorities and the industry on necessary requisites for an adequate and steady supply of aggregates to the construction industry at the optimum balance of social, environmental and economic costs. It is currently under revision.

Department of the Environment (1989) Minerals Planning Guidance 7:
The Reclamation of Mineral Workings, HMSO, London.
This provides advice on reclamation and restoration, which again is an important environmental consideration for the minerals industry.

Department of the Environment (1991) Minerals Planning Guidance 9:
Interim Development Order (IDO) Permissions – Conditions, HMSO, London. Some quarries are working under IDO permissions for mineral working. These are very old permissions granted before the Town And Country Planning Act 1947. Under the recently introduced Planning and Land Compensation Bill there are moves to upgrade the IDO permissions including imposing conditions on environmental impacts. The ways this might be done are given in MPG 9.

Department of the Environment (1988) Minerals Planning Guidance 10:
Provision of Raw Materials for the Cement Industry, HMSO, London.
This gives, among other things, an outline of the specific environmental impacts of the cement industry including the extractive impacts and gives guidance to MPAs on the factors which they should take into account.

Department of the Environment (1988) Planning Policy Guidance 1:
General Policy and Principles, HMSO, London.
This states general planning policy principles and provides guidance on planning procedures.

Box 8.2 (continued)

> Department of the Environment (1988) Planning Policy Guidance 2:
> *Green Belts*, HMSO, London.
> This states the planning policies to be considered in deciding planning applications of developments that may be located in green belts. It states that the extraction of minerals need not be incompatible with green belt objectives provided that high environmental standards for working and landscaping are maintained and that the site is well restored to an appropriate use.
>
> Department of the Environment (1988) Planning Policy Guidance 7:
> *The Countryside and the Rural Economy*, HMSO, London.
> This document describes the planning system in the context of the rural environment by providing a mechanism for balancing the requirements of development and the continuing need to protect the countryside.
>
> Department of the Environment (1988) Planning Policy Guidance 12:
> *Local Plans, Development Plans and Regional Planning Guidance*, HMSO, London.
> This document requires county councils to produce local plans on minerals. The plan must take account of environmental considerations and include policies on the conservation and improvement of the local environment.
>
> Department of the Environment (1988) Planning Policy Guidance 16:
> *Archaeology and Planning*, HMSO, London.
> This states policies on archeological remains and also what weight should be given to them in planning decisions. Again, this is a matter of key importance for the minerals industry.
>
> Department of the Environment (1991) Planning Policy Guidance 18:
> *Enforcing Planning Control*, HMSO, London.
> This gives general advice on enforcement. It discusses what should happen if development goes ahead without permission. It states that there are no special rules for the minerals industry but that the irremediable nature of minerals extraction can pose special problems which may have to be solved by negotiation.
>
> Department of the Environment (1988) Town and Country Planning (Assessment of Environmental Effects, Regulations 1988 SI No 1190, HMSO, London
>
> Department of the Environment (1989) *Environmental Assessment: A Guide to the Procedures*, HMSO, London
>
> Commission of the European Communities (1985) Council Directive of 27 June 1985 on the Assessment of the Effect of Certain Public and Private Projects on the Environment (85/337/EEC). *Official Journal of the European Communities* No L175/40-48 5.7.85.

8.2.3 Control of existing environmental impacts

Under the Town and Country Planning Act 1990 and the Planning and Compensation Act 1991, the Local Government Planning and Land Act 1980 and the Town and Country (Minerals) Planning Act 1981, mineral planning authorities have the power to impose conditions that may govern methods of working, annual output, hours worked, length of quarry life, after use and environmental controls for dust, noise, traffic, etc. The working of a quarry is influenced by these conditions but is also influenced by the Mines and Quarries Act 1954 and the Mines and Quarries (Tips) Act 1969. The mines and quarries legislation is designed to ensure the safety of people employed in the extractive industry and also the general public. It is the responsibility of the Quarries Inspectorate of the HSE to ensure that the legislation is adhered to. The Planning and Land Compensation Act 1991 has introduced important changes to the minerals planning system including the introduction of a new development-plan-led system, and places a statutory requirement on minerals planning authorities to produce local plans.

As well as minerals winning, there is often minerals processing on site. This can often have environmental implications, particularly for air quality. Under the Environmental Protection Act, the Environmental Protection (Prescribed Processes and Substances) Regulations 1991 No. 472 have been introduced to cover major emissions to air, land and water from the most polluting and complex industrial processes, the Part A processes. Further details are given in Section 8.4.2 under Air quality.

Water discharges are covered by the Water Resources Act. Further details are given in Section 8.4.2 under Hydrogeology and surface water.

8.2.4 Other documents on minerals planning

There has been a history of planning circulars from the DoE which have some relevance to minerals planning. These are listed in Box 8.3.

Box 8.3 Department of the Environment Circulars on planning

General

50/78 Report of the Advisory Committee on Aggregates (Verney)
58/78 Report of the Committee on Planning Control of Mineral Working (Stevens)
22/80 Development Control Policy and Practice (part cancelled by 21/91)
22/83 Town and Country Planning Act 1971, Planning Gain
9/84 Hazardous Development
22/84 Memorandum on Structure and Local Plans
1/85 The Use of Conditions in Planning Permissions
30/85 Local Government Act; 1985 Sections 3, 4 and Schedule 1: Town and Country Planning, Transitional Matters
6/86 Local Government (Access to Information) Act 1985
11/87 Town and Country Planning (Appeals) Written Representative Procedure Regulations 1987
1/88 Planning Policy Guidance and Minerals Planning Guidance
3/88 Local Government Act 1985: Unitary Development Plans
10/88 The Town and Country Planning (Inquiries Procedure) Rules 1988, The Town and Country Planning Appeals (determination by Inspectors) (Inquiries Procedure) Rules 1988
10/88 Annex 1 – Code of Practice on preparing for major planning enquiries in England and Wales
10/88 Annex 2 – Code of Practice for hearings into planning appeals
15/88 Environmental Assessment
24/88 Environmental Assessment in Simplified Planning Zones and Enterprise Zones
25/88 General Development Order Consolidation; the Town and Country Planning General Development Order 1988 the Town and Country Planning (Application Regulations) 1988
5/89 The Town and Country Planning (Fees for applications and deemed applications) Regulations 1989
11/91 Control of Pollution (Amendment) Act 1989
14/91 Planning and Compensation Act 1991
16/91 Planning and Compensation Act 1991, Planning Obligations
18/91 Planning and Compensation Act 1991, New Development Plan system
21/91 Planning and Compensation Act 1991: Implementation of the Main Enforcement Provisions

Noise

10/73 Planning and Noise

Waste

55/76 Control and Pollution Act 1974 – Part 1 Waste on Land, Disposal Licences
20/87 Use of Waste Material for Road Fill
13/88 Control of Pollution Act 1974; the Collection and Disposal of Waste Regulations
17/89 Landfill Sites, Development Control
19/91 The Duty of Care

Land use

75/76 Development involving Agricultural land
14/84 Green Belts
28/85 Reclamation and Re-Use of Derelict land

Access

1/83 Public Rights of Way

Employment

14/85 Development and Employment

Surface and groundwater

25/85 Mineral Working - Legal Aspects Relating to Restoration of Sites with a High Water Table
16/89 Water and the Environment
20/90 EC Directive on Protection of Ground Water Against Pollution

Contaminated land

21/87 Development of Contaminated Land

Ecology

32/81 Wildlife and Countryside Act 1981
27/87 Nature Conservation
18/90 Modification to the Definitive Map; Wildlife and Countryside Act 1981
1/92 Planning Controls over SSSIs

8.3 PRINCIPAL ACTIVITIES AND MAIN ENVIRONMENTAL EFFECTS

8.3.1 The scoping exercise

This will be required for all environmental assessments. Further details are given in Chapter 2. The scoping exercise will serve to show what range of skills are necessary to contribute to the environmental assessment. This will vary depending on the type of mineral to be won, the life of the operations, whether the operations are new or an extension of existing operations and whether there will be processing or manufacturing on site such as stone crushing, coal washing and grading, ready mix concrete making, block making or asphalt coating. However, for most projects it is likely that specialists will be required in:

- hydrogeology, geology and hydrology, to determine the effects on the ground and surface water regime

- landscape, to consider the visual impact of the operations and also to assist design of the restoration

- ecology, to assess the effects of species lost or displaced by habitat loss and to help design the restoration to mitigate these effects

- air pollution, noise, and blasting, to predict any potential disturbance to the local population

- traffic, to consider what is often one of the main concerns to people when quarrying operations are proposed

- land use agriculture and amenity, to assess the impact on other land uses, on farmland and on recreation both of the development and any restoration proposals

8.3.2 Baseline data requirements: project activities

All projects will require a (site-specific) description of the site investigation, construction, operational and restoration phases, together with a description of the aspects of the existing environment likely to be affected. The former may include activities such as borehole drilling, top soil removal and storage, construction of bunds for noise and visual reasons, blasting, heavy materials handling, materials processing, waste disposal and the movement of heavy lorries on the surrounding roads. For large quarries with substantial unconsented reserves, several activities may be going on simultaneously on the same site, namely: site investigation, planning applications and environmental assessments, construction activities, minerals winning and processing and restoration. This makes mineral extraction different from many development projects.

Table 8.2 describes some principal activities that may be associated with site investigation, construction, minerals mining and restoration for surface minerals extraction.

Table 8.2 Activities that may result in potentially significant impacts

Phase/activity	Examples	Significance (Significant impact caused) ■ usually □ sometimes
Site investigation/ prospecting		
Transport	Transportation of drilling machinery	□
	Transportation of workers and supplies	□
Site works	Borehole drilling	□
Construction		
Site works	Construction of access roads, site buildings and process plant	□
	Construction of catchment lagoons	□
Transport	Construction of access to local highways	■
	Diversion of rights of way	■
Recreation	Diversion of footpaths and bridle paths	□
	Loss of access to open areas	■
Employment	Employment of local people	□
Land use	Loss of agricultural land, scenery or habitat	■
Operation		
Transport	Removal of materials from site	□
Noise	Heavy earth moving plant, conveyors	■
	Screening	□
	Air overpressure from blasting	■
Vibration	Blasting, transport	■
Lighting	Night-time operations	■
Dust and particulates	Extraction/blasting	■
	Material transport around site	■
	Stockpiling	■
	Removal of materials from site	□
Visual	Position of bunds and stockpiles	■
Employment	Permanent workforce	□
Hydrogeological regime	Effect on surface and groundwater of extraction or dewatering	■
Post operation (some reinstatement ongoing)		
Reinstatement	Landscaping	■
	Planting	■
	Nature reserves	■
	Rights of way and access	■
	Countryside management	■
Land use	Return or land to agriculture or other after-use	■
	Possible site redevelopment	■

8.3.3 Common environmental impacts associated with minerals extraction

Figure 8.3 demonstrates the environmental impacts that may be associated with minerals extraction.

Many of the impacts will be similar for deep mining as for surface mineral extraction, but there are differences, particularly in construction and waste disposal. Minerals processing has a variety of impacts and other sources of aggregate products such as marine dredging have quite different impacts. Therefore, it is not possible within this document to explain every potential interaction of the minerals industry with the environment. Section 8.4 therefore concentrates on what has been shown, from previous environmental studies, to be the impacts that occur most frequently or cause the greatest potential concern to people affected and the relevant consultees.

Site redevelopments which do not include a return to previous use will probably require a separate planning application and possibly an EA. The environmental impact of a redevelopment need not, therefore, be part of a minerals development EA. It must be borne in mind that quarries are sometimes used for landfill and it is the fear of the impacts of this that is often a concern of nearby residents when a minerals planning application is submitted (see also Chapter 9).

8.3.4 Baseline data requirements: environmental parameters

Before a potential environmental impact can be properly calculated it is necessary to have a clear picture of the existing environment. This means collecting high-quality baseline data. The type of baseline data to be collected will be similar for all types of terrestrial minerals extraction and processing. These characteristics are listed in Table 8.3 together with a checklist of common information requirements for baseline data. It should be noted that this checklist is not definitive.

As well as collating existing data, field measurements will almost certainly be required. This is particularly important in respect of the following.

- *Surface and groundwater regime*: An understanding of the hydrogeology in the area of the proposed development will have to be acquired by site investigation directed towards defining aquifers, measuring groundwater levels and describing the flow pattern and the relationship between land drainage and the groundwater regime. This investigation is likely to require the installation of piezometers. Some of this work can be done in association with the reserve assessment and engineering geology feasibility study that will be required for a surface minerals development.

- *Ambient air quality levels*: To determine the effects of dust and fume emissions. The dust emissions would be fugitive emissions from stockpiling and haul roads or from processing such as rock crushing, cement, production, asphalt production, brick making or sand drying. Fumes will arise from mobile plant and process emissions.

- *Ambient noise levels*: To determine the effect of mobile and static plant, particularly important early and late in the working day.

- *Ecological characteristics*: To assess the effects of habitat loss or whether transfer of species is required. Also important for the development of restoration proposals.

- *Geology, geomorphology*: To assess the effect on landform, geological sites and the potential to include geological features in restoration schemes.

- *Land use*: To assess the impact on other land uses, planning, agriculture and amenity, and again to help develop restoration plans.

Figure 8.3 *Typical environmental impacts of extractive processes*

CIRIA Special Publication 96

- *Archaeology*: Any known archaeological sites may have to be investigated further to determine their significance.

- *Leisure, amenity and tourism*: Information on the use of land for recreation, particularly numbers of people using footpaths and open areas both through the site and overlooking the site, is not always available. The quality of the existing country lanes network for walkers, horse riders and cyclists should be recorded.

- *Traffic*: Quarries often bring heavy traffic into rural locations with limited existing facilities, which should be considered in terms of its impact on the local community. The existing road network, traffic flows and accident black spots must be investigated.

Some of this information may need to be collected over an extended period. Groundwater levels, air quality and ecology need, ideally, to be recorded over a year to take account of seasonal variations. Noise and traffic will vary if there are any seasonal influences such as holiday areas, a holiday port or charter flight airport nearby. It is therefore essential that data collection is started at the earliest stage.

Table 8.3 Baseline data requirements

Environmental characteristics	Examples of appropriate data
Soil[1]	Quality, structure, nutrient status, compaction, productivity, erosion
Land use[1,2]	Agriculture, present and potential productivity other land uses. Grades of agricultural land. Derelict land. Forestry. Built environment
Geology and geomorphology[1]	Lithology, structure, location, description of depth and thickness of the seams to be quarried. Location of designated sites
Landscape[1,2]	General character of existing landscape, landscape elements present
Air quality[1]	Particulate levels and fallout, wind direction and speed, rainfall. Existing particulate sources
Noise[1]	Levels L_{A10}, L_{Aeq}, L_{A90}, sources, variability
Vibration[1]	Levels ppv mm/s, sources
Surface water[1]	Quality, drainage, flow, seasonal changes, existing uses and existing drainage
Groundwater[1]	Definition of aquifers, structure and geometry. Identification of recharge and discharge zones. Definition of water table, estimation of seasonal changes. Determination of quality and factors influencing quality
Socio-economics[1,2]	Employment, skills and location of existing workforce, economy, investment, recreation, severance, safety. Residential and other communities
Traffic[1,2]	Existing flows, accident blackspots, current capacity
Flora and fauna[1,2]	Habitat type, rarity, diversity, location of designated sites
Cultural aspects[1,2]	Archaeology on site, architecture/built heritage on site, near site and near access roads

Notes:
1. Survey/assessment maybe necessary.   2. Important sites already well documented.

8.3.5 The role of consultation in the data-collection process

Every effort must also be made to seek out existing data. This is most likely to be available for traffic, ecology, archaeology, geology and hydrogeology, but will be held by a variety of bodies. This illustrates one of the benefits of wide consultations. Some bodies will have to be consulted by the planning authority about proposals. It therefore makes sense to meet with these statutory consultees at an early stage in the EA process. These statutory consultees, plus a list of possible non-statutory consultees, are given in Table 8.4. Further details are provided in Chapter 2.

Consultations should be a significant part of the EA process. The EC Directive 85/337 on EA makes it quite clear that statutory authorities and the public should be consulted during an EA. The Department of the Environment have recently carried out a survey of the way the EA regulations are working in the UK (Wood and Jones, 1991). The conclusion reached by the study was that consultations had not been carried out extensively in all cases, but where they had been, they were generally thought to have been helpful by all sides.

Table 8.4 Possible consultees in the EA process

Statutory consultees include	Non-statutory consultees could include
English Nature	Council for the Protection of Rural England
Countryside Council for Wales	Ministry of Defence
The Countryside Commission	Water companies
Local authority (minerals planning authority)	British Waterways Board
	Sand and Gravel Association
Waste disposal authority	British Aggregate Construction Materials Industries
Local highways authority	Confederation of British Industry
Her Majesty's Inspectorate of Pollution	Local chambers of commerce
The Secretary of State for the Environment	Royal Society for the Protection of Birds
	Parish councils
The Secretary of State for Transport[1]	Local wildlife groups
The British Railways Board[1]	Geological Society
The National Rivers Authority	National Farmers Union
Pollution	Country Landowners Association
Resources	Tenant Farmers Association
Recreation and conservation	Local historical and archaeological societies
English Heritage	Ramblers Association
CADW	Long Distance Walkers Association
County archaeologist	Cyclists' Touring Club
The Theatres Trust	Open Spaces Society
British Coal	British Horse Society
The Ministry of Agriculture Fisheries and Food[1]	Trail Riders Fellowship
	The public, via exhibitions, etc.

Note:

1. In Wales, the Secretary of State for Wales

8.4 IMPACT, IDENTIFICATION, EVALUATION AND MITIGATION

In this section, the various subsections deal in broad terms with identification and prediction of possible impacts. Methods of assessment are then discussed, both by comparison with the existing background and by comparison with absolute standards and criteria. The sources of these absolute standards and criteria are discussed. Finally, mitigation measures for adverse impacts are discussed as are the possibilities for environmental improvements, particularly during restoration. Table 8.5 summarises the impacts covered. Where there are a large number of potential impacts associated with a particular characteristic, then examples of typical impacts are presented in a table.

Table 8.5 Summary of impacts considered

Main category	Impact	Section No.	Table No.
Human impacts	General	8.4.1	
	Noise	8.4.2	8.6
	Vibration	8.4.3	8.7
	Visual impact	8.4.4	8.8
	Traffic and transport	8.4.5	8.9
	Land-use	8.4.6	8.10
	Herigate and archaeology	8.4.7	8.11
Physical and chemical	Air quality	8.4.8	8.12
	Hydrogeology	8.4.9	8.13
	Surface water	8.4.10	8.14
	Geology and geomorphology	8.4.11	8.15
Natural habitats		8.4.12	8.16

8.4.1 Human impacts: general

A recent survey has been carried out for the Department of the Environment on the Control of Noise at Surface Mineral Workings. This included surveys to establish the attitudes of people living near existing or proposed surface mineral workings to the possible environmental impacts, rated in order of significance.

Part of the survey involved a questionnaire sent to local authorities having surface minerals extraction sites in their areas. They were asked what were the main causes of environmental concern for proposed workings and environmental complaint for existing workings. The results are shown in Figure 8.4.

It can be seen that the main causes of complaint were traffic, general noise and dust. Blasting noise and vibration were also significant where blasting occurred. Visual intrusion was a concern at the planning stage but did not generate many complaints.

Comparing proposed and operational responses, noise, traffic and visual intrusion are anticipated as being worse than reality, while the opposite is true for vibration, blasting and, for hard rock quarries, dust.

8.4.2 Noise

Noise can be a significant impact of surface mineral workings. This is shown by the DoE study (DoE, 1990) and the fact that it is also the first environmental impact of mineral working to have a Minerals Planning Guidance Note devoted solely to it (DoE, 1991a). In the introduction to a draft of this Minerals Planning Guidance, Draft MPG11, it is stated that 'The Government recognises that noise from mineral working, while an inevitable consequence of such working, can have a significant impact on the environment and the quality of life of communities'.

The sources of noise in minerals extraction are mainly heavy diesel-powered plant, both mobile and semi-mobile. Various items of fixed plant can also produce noise.

- *Semi-mobile plant:* includes draglines, excavators, loaders and bulldozers working in a restricted area.

- *Mobile plant:* includes scrapers, dumptrucks and graders. A potential problem with any mobile plant is the noise from the vehicle reversing alarms. These are often mentioned specifically when complaints about quarry noise are made by nearby residents.

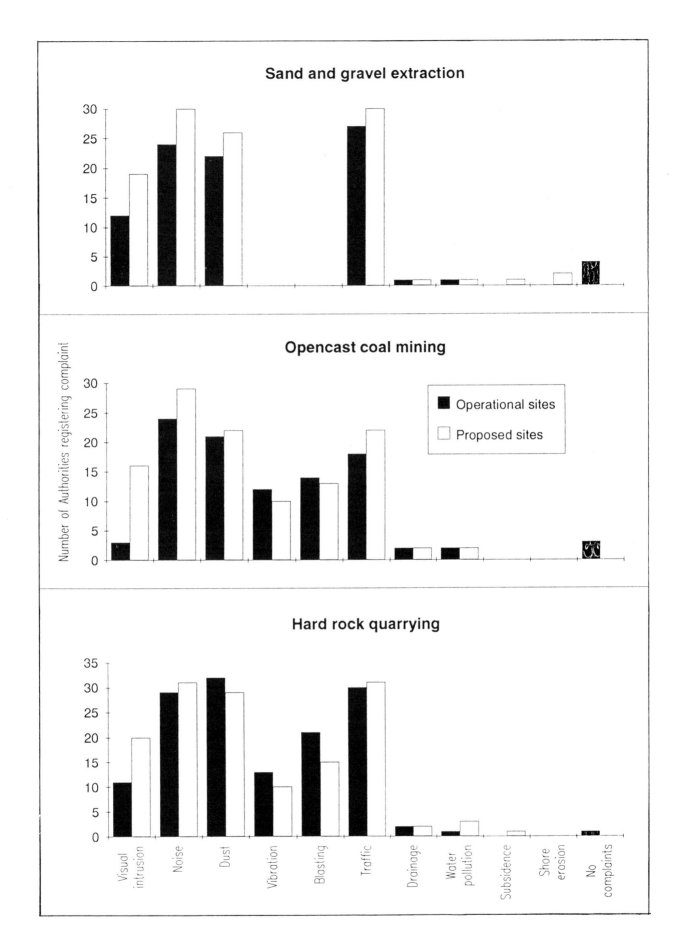

Figure 8.4 *Attitudes of people living near minerals extractions to their environmental effects*

- *Fixed plant:* includes rock crushers, blockmaking equipment, stone sawing and polishing sheds, cement plants, readymix plants, washing drying and screening plant, loading plant and asphalt coating plant. Asphalt coating plants can sometimes cause a noise problem because of the need to start operations vary early in the morning.

Traffic using the site to deliver or pick up products can also cause noise, as can blasting, but these two topics are discussed separately.

Standards for mineral workings have, in the past, been decided on a case-by-case basis, relying on a variety of guidance documents and public inquiry discussions. MPG11, when finally published, should be the main guidance for dealing with noise from surface minerals extraction. No doubt the document will also be referred to in dealing with many aspects of noise from deep mines. Noise from minerals-processing plant separate from extractive sites will be dealt with by reference to DoE Circular 10/73, which is at present also being revised, and a draft replacement Planning Policy Guidance PPGX has been issued (DoE, 1991b).

Draft MPG11 recommends that prediction of noise be carried out largely according to the procedures laid down in BS 5228 (BSI, 1984-6). Some additional advice is given on incorporating soft ground attenuation and meteorological effects.

Table 8.6 Examples of the impact of minerals extraction on noise levels

Sources of impact	Prediction methods	Evaluation criteria	Mitigation options
Construction and operation			
Mobile plant Fixed plant Blasting air overpressure Vehicles	Computer modelling using in house data, manufacturer's data and BS 5228	Draft MPG 11 County MPA Guidelines DoE Report on Control of Noise from Mineral Workings	Alternative plant, e.g. conveyors Screening bunds/fences Local liaison for blasting

There are two methods of assessing the impact of a new noise on an area. The first is to compare it with the existing noise in the area. The second is to compare it with absolute levels that are judged to be satisfactory. Since most minerals extractions are in rural locations with a low background noise level and minerals have to be worked where they are found, there will probably inevitably continue to be discussion over appropriate methods of assessing noise impact and decisions will have to be reached on a case-by-case basis.

There are several mitigating measures that can be used to reduce noise from minerals working. For extraction, much can be done in the way of sequencing and site layout. The works should be sequenced to keep the working face as a barrier between noisy plant and nearby noise-sensitive locations. Noisy fixed plant should be located away from noise-sensitive boundaries, as should haul routes. Baffle mounds or fencing can be used to screen noisy operations. Haul roads should be kept as smooth and well graded as possible. Noise output should be among the criteria considered in selecting new plant and any manufacturer's extra silencing equipment should be considered.

For fixed plant involving air-handling fans, appropriately designed silencers can be used.

Operating hours will also have a significant effect on noise impact and consideration should be given to controlling the times of day at which noisy operations occur. The DoE study (DoE, 1990) showed that setting appropriate planning conditions on noise was generally successful in ensuring acceptable noise levels for local people.

8.4.3 Vibration

Table 8.7 Examples of the impact of minerals extraction from blasting and vibration

Sources of impact	Prediction methods	Evaluation criteria	Mitigation options
Construction For deep mines, blasting for shaft excavation **Operation** For surface mines and quarries, blasting for minerals winning	Computer modelling. Extensive empirical data from US Bureau of Mines and UK data	Human response/ nuisance: BS 6472:1992 Building damage: BRE Digest 353 DIN 4150	Reducing charge weight per delay. Good blast design to maximise energy moving rock and minimise energy going into ground

Some quarries have to use explosives to assist with the mining of minerals. This gives rise to environmental noise and vibration. The explosive power that is used in a particular blast will vary between quarries and will depend on whether the aim is to produce blocks of stone for dimension stone products or whether the mineral is needed in a crushed form.

Modern blasting technology is such that, in many quarries, the vibration and noise from blasting will be clearly perceptible in the environment surrounding the quarry. However, because blasts occur infrequently and at well-defined times, it is possible to give warnings of blasts that will reduce the 'startle factor'.

Prediction of vibration from blasts is difficult. Some of the most comprehensive guidelines are given in two documents by the US Bureau of Mines (Siskind *et al.*, 1980a, 1980b). These present measured vibration and air overpressure levels from a number of blasts in a variety of types of ground. For ground vibration, the data on variation of vibration with distance are presented using the concept of scaled distance. It has been shown that there is an approximately linear relationship between the logarithm of the peak particle velocity at a distance and the scaled distance for the blast. The scaled distance is given by the distance from the blast divided by the square root of the charge weight per delay for the blast. Delays are described further under mitigating measures below.

Predicting air overpressure is particularly complex. The Meteorological Office developed a computer model (Suggitt, 1978) which, given the appropriate detailed meteorological information, can predict ground levels of air overpressure from blasts and, in particular, where focusing might occur. This model was developed mainly for use at weapons-testing sites. Using the computer model merely indicates how sensitive the position of the point of maximum focus is to the conditions in the atmosphere.

Vibration from blasting can disturb in its own right, but it can also cause people concern that their buildings are being damaged. Guidance on annoyance from vibration can be found in BS 6472: 1992 (BSI, 1992). Guidance on building damage risk can be found in BRE Digest 353 (BRE, 1990). This refers to a German Standard DIN 4150 (DIN, 1986) which is widely used in the UK to assess damage risk. What constitutes a satisfactory level of air overpressure is discussed fully in a US Bureau of Mines publication (Siskind *et al.*, 1980b).

Another problem that can arise with blasting is that of flyrock. This is discused briefly in a DoE report (DoE, 1991c). It is stated that problems of flyrock occur about 20 times per year and half of these projections reach 200 m beyond the boundary. Damage is reported to occur in about one third of cases and injury in about 5%.

Mitigation measures for noise, vibration and flyrock centre mainly on blast design and the skill of the blasting contractor. Air overpressure can be reduced by correct stemming and the use of blasting mats. It is also very common for operators to monitor noise and vibration from every blast. As well as this being a good incentive to get the blast design right, it can also be useful evidence to counter claims of damage from blasting vibration or overpressure.

Peak vibration levels can be reduced by arranging small delays between fixing different parts of the blast. These delays are typically 20 ms and do not reduce the blasting efficiency. However, it has been shown that, normally, peak vibrations are largely proportional to the charge weight per delay. Good blast design will ensure that as much of the explosive energy from the blast as possible goes into moving rock and not into the ground.

8.4.4 Visual impacts

Table 8.8 Examples of the visual impact of minerals extraction

Sources of impact	Prediction methods	Evaluation criteria	Mitigation options
Construction and operation			
Working face. Fixed plant and buildings. Mobile plant. Lighting	Plans showing zones of influence. Photomontages and artists impressions showing existing and proposed views	Numbers of sensitive locations having view of works. Subjective assessment of visual quality	Screening bunds and vegetation. Sensitive design of site buildings. Working in directions which minimise views into works. Turning off unnecessary lights at night

The visual intrusion of a quarry depends first of all on the locations from which the quarry can be seen. This will include dwellings and, especially, public vantage points such as roads, footpaths and other rights of way. Quarries on the sides of hills tend to have more impact than those located on a plateau. The principal visual features associated with a quarry will be the working face itself, worked-out and partially restored areas, site offices and buildings associated with material processing and mobile plant on haul roads.

The significance of the visual impact will depend also on the surrounding landscape type and value.

All these features are difficult to measure quantitatively. The main aim in any quarry development that seeks to minimise visual intrusion is to shield as much of the site as possible with natural-looking features such as grassed bunds or existing landscape features. Where funds permit, site offices can be built in local materials and made architecturally attractive. There are no quantitative methods of landscape assessment. Attempts can be made to assess impacts at the design stage through the use of photomontages/artists impressions which are often prepared with the help of computers (see also Section 7.8).

8.4.5 Traffic and transport

Table 8.9 Examples of the impact of minerals extraction on traffic and transport

Sources of impact	Prediction methods	Evaluation criteria	Mitigation options
Construction and operation			
Delivery and removal of product. Workforce	Prediction of traffic flows based on locations of markets for products and locations of sources of labour	DoT *Design Manual for Roads and Bridges*	Limit heavy vehicles to certain routes. Provide on-site overnight parking. Require lorries to be sheeted. Safe design of quarry access

Traffic creates noise and dust as well as affecting the existing traffic flows. Traffic associated with quarrying tends to be mainly large lorries, which can be very noticeable by people living near the quarry.

Noise impacts from traffic can be assessed by considering the changes in average noise level in the working day (usually small) or the changes in peak noise level due to individual pass-bys (usually more significant). Safety at quarry exits due to vehicle movements can be assessed using Department of Transport guidance.

Mitigation measures for traffic can include ensuring that heavy vehicle traffic takes the most environmentally suitable route to and from the quarry. Lorries can be sheeted. On-site parking can be provided to avoid public parking of vehicles overnight. The hours during which traffic enters and leaves the site can be carefully controlled since it is traffic movements out of normal hours that often cause greatest impact. Removal of material from site by rail is sometimes recommended because of its lower impact (DoE, 1991c), although this is only possible in a limited range of cases.

8.4.6 Land use

Table 8.10 Examples of the impact of minerals extraction on land use

Sources of impact	Prediction methods	Evaluation criteria	Mitigation options
Construction and operation			
Loss of agricultural, forestry, heathland, other ecologically valuable land or recreational land from working area. Severance of footpaths. Effects on adjoining land uses and amenity	Estimates of phased land taken out of existing use through working life to restoration and disposition of land uses	Agricultural land classification system, MAFF. Value of habitats and scenic quality. Subjective. Planning policy	Improvements to agricultural land or new recreation facilities during progressive restoration. Final reinstatement

The existing land uses will, of course be affected by quarrying operations. There may be a loss of agricultural or forestry land, although with progressive restoration it may be possible, in certain circumstances, to achieve an overall upgrade of the land in the long term. If neglected or derelict land is included in the extraction area, an overall improvement in the land usage will usually result when extraction operations are complete. However, the aim is usually to restore agricultural land so that there is no loss in high-quality agricultural land. Further guidance is given in MPG 7 and PPG 7 (see Box 8.2). Many quarry operators appreciate the importance of good restoration in making their case for future extraction permissions. The quarrying trade associations actively encourage worthwhile restoration schemes.

The same principles apply to loss of land for recreational purposes. Actual impacts and restoration measures will be very site specific.

Box 8.4 Case study: Regrading of land

		Areas of land (hectares)	
		Existing	Restored
Agricultural	Grades 2 and 3a	9.9	10.1
	Grades 3b	21.6	13.1
	Total agricultural area	31.5	23.2
Non-agricultural — woodland, conservation area, amenity planting, open water		3.5	11.8*
	Total area	35.0	35.0
*3.2 hectare undisturbed, 8.6 hectare new			

Note: There is no longer a Grade 3c in the current Agricultural Land Classification

For instance, at a proposed sand and gravel extraction in Essex, workings would have meant a 14-year extraction operation on 35 ha of agricultural land. The proposals were put forward before the current attitudes towards agricultural overproduction had become widespread. The working operations were carefully phased to minimise the amount of land taken out of production. The existing topsoil was to be stripped in the drier summer months and used to construct screening bunds. The reinstatement proposals reintroduced the soils and regraded the land so that the following improvements were achieved.

8.4.7 Heritage and archaeology

Table 8.11 Examples of the impact of minerals extraction on heritage and archaeology

Sources of impact	Predictive methods	Evaluation criteria	Mitigation options
Construction and operation			
Destruction during excavations. Damage from blasting vibration or dewatering affecting ground stability	Subjective assessment	PPG 16 CBI Code of Practice	Monitor vibration and ground settlement Avoid sensitive sites Funding of rescue digs

Minerals extraction operations may damage or destroy sites of historic archaeological interest. Planning Policy Guidance Note 16 *Archaeology and Planning* and the CBI Code of Practice for Minerals Operators underline the importance of identifying the presence and significance of sites likely to be affected. As a result of consultations, the county archaeologist is empowered to insist on the excavation and archiving of sites of potential interest at the expense of the mineral operator. The expense involved may result in the areas affected being left unworked. Where remains are scheduled under the Ancient Monuments and Archaeological Areas Act 1979 the Secretary of State for the Environment's consent is required before works can proceed. Many quarry companies have funded archaeological rescue digs as a means of allowing eventual winning of minerals or to record unexpected findings made during working.

Damage to ancient monuments, archaeological remains or listed and valued buildings (or their settings) can occur by direct destruction during minerals winning. Damage to ancient buildings surrounding a site may also occur because of blasting vibrations or changes to the groundwater regime causing settlement. Heavy vehicles on surrounding roads may also cause vibration. Recent surveys of traffic induced vibrations on heritage buildings (Watts, 1988; 1989) have concluded that traffic vibrations are not a major cause of damage to heritage buildings.

8.4.8 Air quality

Table 8.12 Examples of the impact of minerals extraction on air quality

Sources of impact	Prediction methods	Evaluation criteria	Mitigation options
Construction and operation			
Earth moving, blasting, stockpiling and bund construction. Minerals moving, crushing, screening stockpiling and removal from site. Traffic on haul roads. Secondary processing such as concrete making and asphalt coating	Computer modelling	DoE and HMIP guidance issued under the Environmental Protection Act	Damping down Grassing of bunds and general perimeter vegetation Enclosing/protecting stockpiles Surfacing and cleaning of haul roads Sheeting of lorries, wheel wash at exits Arrestment systems on processing plant Covering conveyors

Dust from extraction operations was shown by the DoE report (DoE, 1990) to be a principal cause of concern for residents living near quarrying operations. Although dust is the principal air pollutant associated with the minerals extraction industry, vehicles and static plant will produce some fumes, as will processing plant.

Operations likely to produce dust are blasting operations, on-site lorry movements and rock handling. Dust can be generated from wind whipping of stockpiles and quarry floors. Overburden removal and the construction, maintenance and extension of bunds around the site can also cause dust. Processing plant such as crushers, screens, hoppers and associated conveyors can create dust, as can overburden stripping, storage and replacement.

There are no nationally agreed methods of predicting the effects on the surrounding area of dust or gaseous emissions from minerals extraction operations. Various computer models have been used for quarry planning applications and environmental assessments. Models designed primarily to predict airborne emissions from new roads can be adapted to give an indication of particulate emissions from vehicles travelling on dusty roads.

There are no nationally agreed limits for acceptable dust deposition rates. Suitable limits have tended to be established for each quarry at the planning stage, taking into account existing dust levels.

The Environmental Protection Act Part 1 imposes control of emissions to atmosphere, water and land from certain processes including quarries and minerals processing. A number of guidance notes have been issued (see Box 8.5). Those written for Local Authority enforcement — the Secretary of State's Guidance — contain advice on mitigating the effects of dust. They contain the following possible measures:

- covered storage of stocks of finely crushed rock (3 mm and under)
- covered conveyors
- dust-tight screen houses
- dust-extraction and filtering on crushers
- roadgoing vehicles to be sheeted
- roadways to be hard surfaced and kept clean
- vehicle exhausts not to point down
- wheel-washing facilities at the entrance.

Box 8.5 Environmental Protection Act guidance

Secretary of State's Guidance (DoE)

PG3/1(91)	- blending, packing, loading and use of bulk cement
PG3/2(91)	- manufacture of heavy clay goods and refractory goods
PG3/5(91)	- coal, coke and coal product processes
PG3/7(91)	- exfoliation of vermiculite and expansion of perlite
PG3/8(91)	- quarry processes including roadstone plants and the size reduction of bricks, tiles and concrete
PG3/9(91)	- sand drying and cooling
PG3/10(91)	- china and ball clay
PG3/11(91)	- spray drying of ceramic materials
PG3/12(91)	- plaster processes
PG3/14(91)	- lime slaking processes

Chief Inspector's Guidance to Inspectors (HMIP)
IPR3 Minerals Industry Sector

8.4.9 Hydrogeology

Table 8.13 Examples of the impact of minerals extraction on hydrogeology

Sources of impact	Prediction methods	Evaluation criteria	Mitigation options
Construction and operation			
Loss of natural filtration of rain water Destruction of aquifers Mine drainage water pollution Dewatering	Computer modelling	Water Resources Act NRA Requirements	Leachate lagooning. Grout curtains. Avoid aquifers. Work minerals wet or leave margin Store water for replenishment

Many on- and off-site activities may affect underground water quality and quantity. This can include such things as contamination from quarry drainage water, road deicing chemicals, chemical dust suppressants, pesticides and fertilisers.

Alterations to the water table and groundwater can also be significant. Judith Williams (1991) reports the extensive discussions on quarrying which have occurred in the Mendips. Quarrying has been carried out in the Mendips since 1870 and to date 250 million tonnes of stone have been removed. The Mendip aquifers produced more than 90,000 million litres of drinking water each year and the rock above the water table acts as a temporary storage and filtration zone for the rainwater falling on the hills. The NRA has been concerned that if too much stone was quarried this filtration effect would be reduced, resulting in the extracted groundwater being more polluted. The quarrying companies have tackled the problem by carrying out detailed hydrogeological studies and incorporating water storage reservoirs as part of the quarrying operation to augment water supplies, particularly in the summer. The hydrogeology of the Mendips has been the subject of a British Geological Survey Report (BGS, 1992).

This illustrates that such hydrogeological effects must be an important consideration of any quarry development. Ellison (1988) reports that repeated shock waves caused by blasting can create new fractures in rock and may lead to damage to shallow aquifers and rearrangements of local groundwater flow.

Any minerals extraction which requires dewatering will give a risk of settlement of nearby buildings if these are built on granular material from which water is removed. This can be particularly significant where clay and granular material lie in bands.

Predictions of the effects of quarrying operations on the groundwater table will require input from the engineering designers of any projects and an expert in hydrogeology. Assessments of the effects on local water quality and ground settlement will be very site specific and must again be carried out in consultation with a hydrogeology specialist. The effects on flora, fauna and agriculture will require input from an ecologist. Mitigating measures are likely to be very site specific but could include the following.

- Arrange to collect waste water from operations and treat this before it is disposed off site. This collection could be done with grout curtains to constrain water to collect in lagoons.

- Minerals extraction operations could be programmed to avoid periods of high rainfall.

- The restoration proposals should consider how to restore the water table without the possibility of groundwater pollution.

- Roads should be surfaced and bunds should be grassed to minimise siltation.

- Dewatering can be done in cells, adequate margins can be left from sensitive sites or dewatering can be avoided by working minerals wet.

8.4.10 Surface water

Table 8.14 Examples of the impact of minerals extraction on surface water

Sources of impact	Prediction methods	Evaluation criteria	Mitigation options
Construction and operation			
Changes in water flows Siltration. Chemical contamination	Modelling	Water Resources Act NRA Requirements	Leachate lagooning Divert water courses around workings. Grass bunds

Minerals extraction may affect the flows of surface water courses where these cross the area of working. When an area is devegetated and overburden removed, surface water runoff can be greater and quicker, particularly in storms. This runoff water can contain suspended solids or chemical contaminants. This can affect the water quality and the aquatic flora and fauna of nearby watercourses. Chemical contamination can arise from soluble chemicals washed out of the minerals being worked as well as chemicals brought on to the site such as fuel oil and deicing chemicals.

To minimise problems, the existing water flow patterns and water quality in the area of proposed minerals working should be established at the planning stage. The aim should be to preserve these flow patterns and water qualities during the execution of the workings. Water courses can be diverted around the workings and kept separate from quarry water by bunding. These bunds and storage bunds for overburden should be grassed to minimise siltation. Lagoons can be constructed to prevent excessive runoff in storms and to allow sediment to settle out. If chemical contamination is a problem the lagoon can collect water prior to any necessary chemical treatment. When minerals are worked on a flood plain the storing of overburden or waste on the flood plain should be avoided. If this is not possible, then stores should be aligned so that they provide the minimum impedance to flood flow.

8.4.11 Geology and geomorphology

Table 8.15 Examples of the impact of minerals extraction on geology and geomorphology

Sources of impact	Prediction methods	Evaluation criteria	Mitigation options
Construction and operation			
Destruction of sites of geological or geomorphological interest	Subjective assessment	Wildlife and Countryside Act	Avoid sensitive areas Creation of new geological exposures to be incorporated in restoration

Minerals extraction can affect geological and geomorphological features through their direct destruction or altering the natural processes affecting such sites. New quarries are often proposed where there has been a history of small-scale exploitation in the past. These small old quarry sites may now be locations where valuable geological information can be gained and some may be of SSSI status.

Mitigation measures can include the avoidance of sensitive areas. There is also the potential to create new geological exposures for incorporation on restoration plans.

8.4.12 Natural habitats

Table 8.16 Examples of the impact of minerals extraction on natural habitat

Sources of impact	Prediction methods	Evaluation criteria	Mitigation options
Construction and operation			
Destruction of habitats and flora. Changes to water regime. Dust and blasting	Subjective assessment	Wildlife and Countryside Act	Avoid sensitive areas Species transfer Construction of new habitats during progressive restoration

Minerals extraction can effect local ecology directly through destruction of flora and habitats for fauna. There will also be indirect ecological effects from pollution from the site affecting surrounding areas, particularly dust pollution and startle from blasting. Increases in vehicle movements on local roads may also have an effect. As well as destruction of habitats of fauna permanently resident in the area, migration routes may be disrupted. The restoration phase of minerals working gives the opportunity to create new habitats and to restore species and migration routes. Sediments and leachates from the site can affect local watercourses and aquatic flora and fauna. Dewatering can affect aquatic habitats directly and other flora by changing the soil conditions. This applies particularly to bog or mire areas. Recharging by pumping may not be appropriate because of changes in the water quality (DoE, 1991c).

Mitigating measures can include avoiding minerals mining in areas of high ecological value and maintaining them within the site boundary. Nature reserves and extraction can sometimes coexist. Rare or endangered species can be transplanted from the site before work begins. Soil stripping or replacement can be timed to avoid the breeding season. Hedges can be transplanted but not during nesting seasons. Margins can be left around or along trees and hedges. Restoration gives the opportunity to reintroduce rare species and to develop new habitats to increase species diversification. The local wildlife trust can be involved in the design and management of restored areas. In order for the above mitigating measures to be effective there must be an extensive ecological baseline survey before any work begins. Further monitoring as the works progress can then assess the effectiveness of the mitigating measures and identify any species, particularly rare or endangered that may have been missed during the initial survey (see also Section 7.12).

8.5 REFERENCES

ANON (1979)
Ancient Monuments and Archaeological Areas Act
HMSO (London)

BACMI (1992)
Environmental Code
British Aggregate Construction Materials Industries (London)

BRE (1990)
Damage to structures from ground borne vibration
Digest 353, Building Research Establishment (Garston)

BRITISH GEOLOGICAL SURVEY (1992)
Limestone resources and hydrogeology of the Mendip Hills
Technical Report WA/92/19, BGS (London)

BROOKE, J.S. and WHITTLE, I.R. (1990)
The role of environmental assessment in the design and construction of flood defence works
IWEM90 Conference (Glasgow)

BSI (1984-6)
Noise control on construction and open sites
BS5288 Parts 1-4, British Standards Institution (London)

BSI (1992)
Evaluation of human exposure to vibration in buildings (1Hz to 80Hz)
BS6472:1992, British Standards Institution (London)

CBI
Archaeological investigations – Code of Practice for minerals operators
Confederation of British Industry (London)

DIN (1986)
Structural vibrations in buildings, Part 3 Effects on structures
DIN4150, Deutches Institut fur Normung (Berlin)

DoE (1990)
The control of noise at surface minerals workings
HMSO (London)

DoE (1991a)
The control of noise at surface mineral workings
Draft MPG11 dated 3.10.91, Department of the Environment (London)

DoE (1991b)
Planning Policy Guidance, *Planning and Noise*
Consultation Draft PPGX, date 31.12.91, Department of the Environment (London)

DoE (1991c)
Environmental effects of surface mineral workings
Department of the Environment Research Report, HMSO, (London)

ELLISON, R.L. (1988)
Review of coal strip mining EA: The accuracy of an analysis
Environmental Impact Assessment Review Vol 8 pp 221–232

GROUNDWORK (1991)
The minerals industry environmental performance study
Groundwork Associates Limited (Knutsford)

SAND AND GRAVEL ASSOCIATION (1991)
Code of practice
The Sand and Gravel Association, (London)

SISKIND, D.E. *et al.* (1980a)
Structure response and damage produced by ground vibration from surface blast mining
RI8507 United States Department of the Interior (Washington)

SISKIND, D.E. *et al.* (1980b)
Structural response and damage produced by airblast from surface mining
RI8485 United States Department of the Interior (Washington)

SUGGITT, R.T. (1978)
A method for the production of acoustic forecasts using a digital computer
Meteorological Magazine, Vol 107, pp 374–383

WATTS, G.R. (1988)
Case studies of the effects of traffic induced vibrations on heritage buildings
Research Report 156. Transport and Road Research Laboratory (Crowthorne)

WATTS, G.R. (1989)
The effects of traffic induced vibrations on heritage buildings – further case studies
Research Report 156. Transport and Road Research Laboratory (Crowthorne)

WILLIAMS, J. (1991)
Quarrying – Fresh problems springing up
Construction News Supplement issued June 1991 (London)

WOOD, C. and JONES, C. (1991)
Monitoring environmental assessment and planning
Department of the Environment Report, HMSO (London)

CIRIA Research Project 424

Environmental Assessment

A guide to the identification and mitigation of environmental issues in construction schemes

9. Waste management

Jane Barron, Principal Consultant, W S Atkins Environment

9 Waste management

9.1 TYPES OF SCHEME CONSIDERED

9.1.1 Waste arisings

The waste industry is in the throes of change. Legislative, economic and environmental pressures are forcing technological improvements in the traditional areas of landfill and incineration and encouraging the development of new methods of dealing with wastes. In certain areas, strategies are being implemented that will ultimately lead to zero landfill disposal for controlled wastes (defined below). Many of the projects and schemes aimed at minimising the volumes of waste for ultimate disposal require the design and construction of new plant and there is currently considerable activity in this sector. Although a reduction in the use of landfill will inevitably reduce the associated environmental disbenefits, landfill frequently has a major benefit in the reclamation of derelict land (particularly old quarry workings), which will then have to be realised by different means.

The major proportion of all arisings of waste come from agricultural sources (250 million tonnes in 1990) with a further high proportion from mining and quarrying sources (108 million tonnes in 1990). Of the 516 million tonnes of waste arising in the UK in 1990 (figures taken from Waste Management Paper No.1, 1992) only some 137 million tonnes were controlled wastes, and of these it is estimated that some 27 million tonnes (1986/87 figure) were handled by English and Welsh waste disposal authorities. About half of this 27 million tonnes was received from collection authorities, nearly one third from commerce and industry, about one sixth from household (i.e. civic) amenity sites and the rest from miscellaneous sources.

Controlled wastes comprise household wastes, commercial wastes, industrial, demolition and construction wastes and sewage sludge when landfilled or incinerated but not otherwise. Neither agricultural nor mining and quarrying wastes are regulated as controlled wastes under the Control of Pollution Act 1974 or the Environmental Protection Act 1990.

This chapter deals with the environmental effects arising from the construction and operation of waste treatment, transport and disposal facilities for controlled wastes.

9.1.2 Waste management methods and environmental protection

With regard to disposal of controlled wastes, by far the majority, i.e. about 97%, is disposed of to landfill, approximately one third to local authority sites (or, following privatisation, former local authority sites) and the remainder to private sector sites. Consequently much of this chapter is given over to discussion of effects arising from the disposal of controlled wastes to landfill.

However, the treatment of wastes is becoming increasingly important, both to meet the requirement that up to 50% of the recyclable portion of domestic waste (i.e. 25% of waste) be recycled by the year 2000, and to reduce the bulk for both transportation and disposal. This both reduces the cost of transport of waste materials and increases the length of life of landfill sites. In addition, in the UK there is increasing interest in the recovery of energy from waste and in the use of waste for composting to return nutrients to the land.

The government's White Paper *This Common Inheritance* encourages the reuse or recycling of materials that otherwise would be thrown away and, in default, aims to recover energy from those wastes that either cannot be recycled or for which it is uneconomic to do so.

The environmental effects of waste collection, treatment and disposal processes depend to a large extent on the type of waste being disposed of, the quantity and the method of transport used. Changes in policy and legislation can be expected to have an effect on these parameters.

Environmental awareness and economic pressures at a commercial level have been reflected at a political level by the introduction, in the Environmental Protection Act 1990, of regulations requiring the minimisation and recycling of waste. Various initiatives are being put in place to support these regulations. The government is in particular very keen to pursue the use of construction materials as secondary aggregates, which will both limit the environmental effect of their transport, treatment and disposal and minimise the effect of the alternative of mineral extraction.

Box 9.1 Waste reduction at source: reuse and recycling

In many cases, the choice of manufacturing or industrial process may result in the utilisation of environmentally favourable waste management strategies, i.e. waste minimisation, re-use, recycling or disposal. Industrial producers and other generators of waste are therefore faced with the need to balance environmental effects and economic considerations. The process of environmental auditing looks at all levels of a company's operation and looks for ways in which cost savings related to efficiency may be implemented; these also relate to recycling, energy conservation and waste minimisation. Life-cycle analysis may cause the environmental effects of waste handling to be reduced. This involves the cradle-to-grave analysis of particular products and their construction with respect to the raw materials used, the energy consumption necessary for manufacture and operation, the possibility of recycling, the need for disposal and release of pollutants to the environment. In certain cases, life-cycle analysis attempts to determine the form of packaging that overall has greatest potential for recycling and the least impact on the environment; equally valuable is analysis to compare the benefits of returnable containers with one-trip disposable ones. The manufacture of goods very often requires the winning and processing of large quantities of raw material and the production of large quantities of wastes. This is another area where attention paid to the scale of the process or the quantity of waste being produced can effectively reduce the environmental effects.

Following on from these methods of minimising and recycling wastes at source is the development of a growing variety of processes for the treatment and recycling of generated wastes ('end of pipe' solutions). Such methods include:

- energy from waste incinerators
- large-scale composting plants
- anaerobic digestion systems for sewage sludge
- waste separation plants and transfer stations
- waste treatment plants (for industrial wastes and sludges)
- contained landfill with engineered liners and caps, and sophisticated monitoring systems
- schemes to utilise landfill gas for power generation.

As described earlier, the majority of wastes are disposed of to landfill – the controlled disposal of waste generally into holes in the ground. These holes may have arisen from mineral extraction, may be disused railway cuttings, marshy areas or general depressions in the ground. There is an increasing prevalence of above-ground landfills, which are sometimes known as landforms. Here the practice is to dispose of waste again in a controlled manner but above the ground to form a hill or mound. The practice of land-raising (i.e. increasing the height of an existing landfill or raising the final profile above existing contours) is proving an economic and increasingly popular method of increasing void space and extending the life of a site. The environmental effects of doing this will often be less than those arising from the development of a new site, and benefits will accrue from the improved surface water runoff and reduction in leachate formation.

The incineration of wastes in the UK has had a somewhat chequered history, initially coming into favour in the late 1960s and early 1970s, falling out of favour in the 1980s and appearing to come back into favour (at least conceptually) in the 1990s. With the cessation of the disposal of wastes to sea in 1993 and disposal of sludge to sea in 1998, there is growing interest in the incineration of wastes, particularly of a combination of organic wastes and the possibility of recovery of energy from this process.

This chapter inevitably concentrates on the environmental effects arising from landfill and incineration. However, where specific environmental effects arise out of the construction, operation, maintenance or demolition of waste treatment plants then these are described.

Box 9.2 Trends and specific problem areas

The direction of movement is apparent but change is being constrained by the recession and the squeeze on public sector spending which is holding back the development of recycling and more innovative solutions to the disposal of domestic waste. Specific aspects which will affect the rate of change and direction of movement in the waste management industry are as follows.

- Continuing tightening of legislation and environmental controls is to be expected.

- Waste minimisation is already showing effects in the industrial sector, and recycling is set to become more important in domestic and commercial sectors. Despite this there is a growing demand for the disposal of special and controlled wastes as standards are tightened and dispersion to atmosphere and watercourses is reduced.

- A wider range of waste disposal solutions is to be expected; demand for specialist treatment facilities is growing; the range of solutions for disposal of domestic and commercial waste is developing including recycling, materials recovery, energy from waste, and composting schemes.

- Landfill is moving towards containment for all putrescible and potentially leachate producing wastes; energy recovery schemes to utilise landfill gas will become the norm.

- With cessation of sea dumping, sewage sludge disposal will become a major waste management issue; incineration and/or anaerobic digestion will be required, involving the creation of major new facilities.

- Special problems like tyres, batteries, clinical waste, animal carcasses, and aerosols will require special solutions.

- Agricultural wastes.

- Contaminated land treatment.

- Fiscal instruments, including landfill levies, carbon tax, financial bonding for aftercare requirements on landfill sites, and recycling credits.

- Liability of closed landfill sites.

Table 9.1 Examples of waste management operations

Scheme type	Typical promoter	Development type	Type of works
Collection and transfer	District authorities Private contractors	Depots Transfer stations	Buildings, infrastructure Vehicle maintenance Plant
Recycling/recovery	Government departments District authorities Private contractors Manufacture and process industries	Depots Recycling plant Energy from waste	Buildings, infrastructure Vehicle maintenance Plant
Treatment	Local authorities LAWDCs[1] Private contractors Manufacture and process industries	Incinerators Balers Composters Digesters Pulverisers Chemical treatment Biological treatment Recovery Physical treatment	Buildings, infrastructure, Plant
Disposal	Local authorities LAWDCs Private contractors Manufacture and process industries	Landfill Reclamation of marginal and derelict land Landform	Earth works Synthetic liners; significant drainage works Some buildings and infrastructure

Note:

1. Local Authority Waste Disposal Company

9.2 KEY STANDARDS AND LEGISLATION

9.2.1 Environmental Assessment requirements

Circular 15/88 giving guidance on the Town and Country Planning (Assessment of Environmental Effects) Regulations 1988 advises that an environmental assessment will be required for any facility for the disposal of controlled waste (landfills) where the quantities of waste to be deposited are more than 75,000 tonnes per year. Incinerators for special wastes fall into a different category under the Environmental Assessment Regulations and an EA is mandatory under these regulations, regardless of capacity. Special wastes are those that are particularly hazardous to human health or the environment and are defined under the Control of Pollution (Special Wastes) Regulations 1980.

All waste treatment or disposal facilities require planning permission and a licence; many also require authorisations and consents to discharge effluents. Planning applications are made under the Town and Country Planning Act 1990 and licences are currently issued under the Control of Pollution Act 1974 (from April 1994 licences are to be issued under the Environmental Protection Act 1990). It is not possible to obtain outline planning permission for landfill sites, nor is it possible to obtain a licence without previously having obtained planning permission. As described above, most facilities will require an Environmental Statement to accompany the planning application, this statement to be prepared under the Town and Country Planning (Assessment of Environmental Effects) Regulations 1988.

The environmental assessment may be volunteered by the developer or requested by a number of authorities:

- the local planning authority
- the Secretary of State for:
  - Environment (Town and Country Planning)
  - Wales (developments in Wales).

There is little formal guidance on the EA of waste disposal facilities. DoE/Welsh Office publication, *Environmental Assessment: A Guide to the Procedures* (1989), gives general guidance as to the issues that must be assessed. A draft document on EA of landfill sites was issued by DoE for consultation in 1989 and this did little other than reiterate the guidance given in the DoE/Welsh Office document. Further guidance can be found in the DoE Circular 15/88 (Welsh Office 23/88). All these documents emphasise that the need for an EA should be determined by the likelihood of significant environmental effects arising and not by the amount of opposition or controversy to which a project gives rise.

9.2.2 Other relevant legislation

Because of the nature of waste management and disposal facilities, there is a very wide range of legislation affecting their planning, development and use. For instance, landfills can have an impact on traffic, groundwater, nature interests, local residential interests and industrial interests, and may give rise to general nuisance. An incinerator may result in air pollution, water pollution and noise, and may also give rise to transportation effects. Relevant UK legislation is listed in Table 9.2; EC legislation and international conventions are listed in Table 9.3.

In addition to the above, the Environmental Protection Act 1990 places a Duty of Care on anyone handling or disposing of waste to ensure that waste is always handled by fit and proper persons, who will have proper regard for environmental protection. The Environmental Protection Act will replace the licensing function in the Control of Pollution Act 1974, and require a Certificate of Completion to be granted by the Regulation Authorities before a licence can be handed back by an operator. The Act will also bring in Integrated Pollution Control, which will have an effect on the characteristics and quantities of waste produced and their environmental effects.

The DoE has produced a series of Waste Management Papers, which give guidance on the management of wastes. Under the Environmental Protection Act 1990 it will be possible to give statutory backing to parts of these papers and it is understood that Waste Management Paper (WMP) No. 27 *Landfill Gas* is likely to be the first to be treated in this way. In addition to WMP 27, WMP 26 (*The Landfilling of Waste*), WMP1 (*Waste Disposal and Treatment — a Review of Options*) and the draft of WMP 4 (*Licensing of Waste Facilities* — currently being revised) are of particular relevance to the environmental assessment of waste management schemes. The full list of Waste Management Papers, including those currently in draft, is given in Table 9.4.

Useful guidance is now being prepared by a number of professional bodies:

- The Institute of Environmental Assessment: Guidelines on Ecological Information in Environmental Statements

- English Nature: Guidelines for developers on the involvement of statutory consultees in the Environmental Assessment process

9.2.3 Responsibilities of controlling agency

Waste management control is dispersed between several agencies. In addition to planning controls (designed principally to protect visual amenity, and to minimise adverse effects on surrounding land use, and public amenity), a sophisticated system of environmental protection is developing for waste management operations involving the National Rivers Authority, Her Majesty's Inspectorate of Pollution (HMIP), local-authority-based waste regulation authorities, and district-council-based environmental health authorities.

Currently the responsibilities break down as follows.

NRA

- Involved in any waste handling that poses a risk of contamination to ground or surface waters.

- Issues consents for discharges from individual processes and waste treatment plants.

- Acts as statutory consultee for waste management/disposal activities that require a licence from the local authority, authorisation from HMIP, or permit from District Environmental Health Officer.

- Recommends conditions for planning consents and waste disposal licences which aim to prevent pollution of controlled waters.

- Tighter controls on permitted effluent release and new guidelines on protecting sensitive areas are having a far-reaching impact on waste management practice. All potentially polluting landfills and industrial/waste management processes should now be 'contained' to prevent egress of runoff or leachate from site, and collected waste waters often require treatment before release to watercourse or sewer, with the result that pollutants are now being concentrated as sludges or solids requiring controlled disposal — often to landfill.

HMIP

- Issues authorisations for combustion plants (except small plant capable of burning less than one tonne per hour) with stipulations to control atmospheric emissions.

- New authorisation procedure requires all processes to come up to new standards under BATNEEC (Best Available Technology Not Entailing Excessive Cost) by 1995. The result is likely to be closure or major remedial works at many plants, including all but a few of the UK's existing waste incinerators. Tighter controls on particulate and organic emissions will require major investment in gas scrubbers and precipitators and in more sophisticated designs of furnace, and more sophisticated control and monitoring systems. The net result will again be the concentration of pollutants previously dispersed to atmosphere into contaminated ash, particulates, and sludges which also require controlled disposal — normally to landfill.

Waste regulation authorities

(County councils in England except in former metropolitan areas where special arrangements apply).

- Responsible for issuing, monitoring, controlling and enforcing all regulations, including licences, for all waste management activities including landfill sites, importation, handling and storage of waste at incinerators, transfer stations and waste handling operations such as materials recovery, waste storage, and scrap yards.

- Licensing powers were increased with the Environmental Protection Act 1990 and are now being implemented; these controls will be tightened further by new legislation including the impending EC Landfill Directive.

- Definitions of 'special waste' are changing with the result that wastes are being moved up the scale in terms of hazard to the environment (i.e. from hazardous or difficult to special) — a higher proportion now requires treatment, incineration, or controlled landfill (including residues from treatment plants and incinerators).

Local authority pollution control (principally air pollution control)

(Based on District Council Environmental Health Department).

- Issues permits for small scale incineration/combustion plants.
- Has some responsibility for general environmental monitoring.

These environmental protection responsibilities are over and above the normal planning control system and somewhat complicate the submission of EAs for waste management projects. The details of industrial or combustion processes are, for example, exempt from consideration in an EA because they are covered by the HMIP Authorisation process, but the National Rivers Authority and the waste regulation authorities are expected to work closely with planning authorities in receiving and dealing with applications for waste treatment and waste management operations.

9.2.4 Public perception

There is a general perception by the public that facilities used for the treatment, transfer or disposal of waste are extremely polluting and a nuisance to their neighbours. While this may have historically been the case, it is not necessarily the case in the 1990s. However, the perception remains, and for this reason an EA may be required to accompany planning applications even for small waste facilities. An unbiased EA will facilitate objective determination of the planning issues and ensure proper environmental protection.

Incineration plants and toxic waste treatment plants have received bad press in the past and are commonly blamed for high incidences of specific diseases in their locality. Consequently, the NIMBY (not in my back yard) syndrome applies; nobody wants a waste facility near to where they live. However, sensitive planning and public consultation can overcome this prejudice (see Section 9.3.5).

Table 9.2 Summary of UK legislation relevant to the EA process

Date	Act/Regulations	Purpose/relevance
1906	Alkali etc. Works Regulations Act 1906	Regulation of emissions to atmosphere; largely superseded by the EPA Act 1990
1960	Radioactive Substances Act 1960	Controls disposal of radioactive substances
1968	Clean Air Act 1968	Control of grit, dust and smoke
1968	Countryside Act 1968	Setting up Countryside Commission, access to the countryside
1972	European Communities Act 1972	Controls vehicle noise
1972 and 1974	Road Traffic Acts 1972 and 1974	Control of noise
1973	Nature Conservancy Council Act 1973	Setting up of the Nature Conservancy Council, regulations
1974	Control of Pollution Act — will be largely superseded by the EPA when relevant sections enacted	Principal statute for control of noise; registration of carriers; control of special waste; waste licensing
1974	Health and Safety at Work Act 1974	Safe working
1981	Wildlife and Countryside Act 1981	Nature conservation; SSSI
1985	Wildlife and Countryside (Amendment) Act 1985	SSSI Notification
1989	Water Act 1989	Setting up of National Rivers Authority; control of water pollution; NRA's environmental and recreational duties
1989	Electricity Act 1989	Obligations with regard to the use of non-fossil fuels
1989	Control of Pollution (Amendment) Act 1989	Registration of carriers
1990	Environmental Protection Act 1990	The control of pollution to water, air and land; waste disposal licensing, duty of care, carriage of waste, statutory nuisances, litter control, recycling, waste minimisation, integrated pollution control
1990	Town and Country Planning Act 1990; Amended by the Planning and Compensation Act 1991	Requirement for planning permission
1991	Water Resources Act 1991	Discharges to Controlled Waters
SI 1980/1709	Control of Pollution (Special Waste) Regulations; and Amendment of 1980	Control of special wastes through the consignment note system
SI 1983/943	Health and Safety (Emissions to the Atmosphere) Regulations 1983	Prevention of the release of noxious or offensive substances
SI 1988/1562	Transfrontier Shipment of Hazardous Waste Regulations 1988	Control over the import and export of waste including labelling, insurance etc.
SI 1988/1657 Amended SI 1991/2431	Control of Substances Hazardous to Health Regulations 1988	Safe working
SI 1991/1624	The Controlled Waste (Registration of Carriers and Seizure of Vehicles) Regulations 1991	Registration and control of carriers of waste
SI 1991/2839	The Environmental Protection (Duty of Care) Regulations 1991	Places a strict duty of care on all producers, brokers, transporters and disposers of waste to ensure its safe disposal and the application of the consignment note system to all wastes
SI 1992/742	The Environmental Protection (Duty of Care) Regulations 1991	Control on the transport of hazardous waste
SI 1992/743	Dangerous Substances (Conveyance by Road in Road Tankers and Tank Containers) Regulations 1992	Control on the transport of hazardous waste

Table 9.3 Summary of EC Directives relevant to the EA process

Date	Convention/Directive	Purpose/relevance
1975	75/439/EEC Directive on the disposal of waste oils	To ensure that the collection and disposal of waste oils causes no avoidable damage to humans and the environment
1978	78/659/EEC Directive on the quality of fresh waters needing protection or improvement in order to support fish life	Protection of freshwaters from contamination by leachate; deals with the quality of water required to support salmonid and cyprinid species of fish
1979	79/409/EEC Directive on the conservation of wild birds	Setting up of Special Protection Areas (SPA) for birds
1980	80/68/EEC Directive on the protection of groundwater against pollution caused by certain dangerous substances	Protection of exploitable groundwaters from contamination by leachate
1982	75/66/EEC Ramsar Convention on the Protection of Wetlands of International Importance	Designation of internationally recognised sites
1982	82/72/EEC Berne Convention on the Conservation of European Wildlife and Natural Habitats	Mandatory requirements to conserve wild flora and fauna and their natural habitats, giving particular emphasis to endangered and vulnerable species
1982	82/461/EEC Bonn Convention on the Conservation of Migratory Species of Wild Animals	Strict protection for certain species in danger of extinction throughout all or a significant portion of their range; agreements for management of other species
1984	84/631/EEC Directive on the supervision and control within the European Community of the transfrontier shipment of hazardous waste	Control on the transfrontier movement of hazardous waste including compliance with requirements of the member states, insurance labelling and instructions in the event of danger or accident
1985	85/337/EEC Directive on the assessment of the effects of certain public and private projects on the environment	Environmental assessment requirements
1989	89/369/EEC Directive on the prevention of air pollution from new municipal waste incineration plants	Fixes emission limit values for new incinerators
1989	89/429/EEC Directive on the reduction of air pollution from existing municipal waste incinceration plants	Fixes emission limit values for existing incinerators
1991	91/156/EEC Draft Framework Directive for Waste; will Supersede 75/442/EEC Directive on waste	Describes definitions of waste and licensing systems
1991	91/689/EEC Directive on Hazardous Waste; Supersedes 78/319/EEC Directive on toxic and dangerous waste	The application of 75/442/EEC to hazardous wastes
1992	Draft Directive on the landfill of Waste — 1st reading of 7th Draft	Design requirements to prevent escape of gas and leachate
1992	Draft Directive on Packaging and Packaging Waste	Reduction of and control of packaging waste
1992	92/43/EEC Directive on the conservation of natural habitats and of wild flora and fauna	Designation of protected habitats

Table 9.4 Waste Management Papers

Number	Title
1	Reclamation, Treatment and Disposal of Waste; 1976, Revised 1992
2	Waste Disposal Surveys; 1976
3	Guidelines for the Preparation of a Waste Disposal Plan; 1976
4	The Licensing of Waste Disposal Sites; 1976
5	The Relationship between Waste Disposal Authorities and Private Industry; 1976
6	Polychlorinated Biphenyl (PCB) Wastes – a Technical Memorandum on Reclamation, Treatment and Disposal; 1976
7	Mineral Oil Wastes – a Technical Memorandum on Arisings, Treatment and Disposal; 1976, Second Edition 1985
8	Heat Treatment Cyanide Wastes – a Technical Memorandum on Arisings, Treatment and Disposal; 1976, Second Edition 1985
9	Halogenated Hydrocarbon Solvent Wastes from Cleaning Processes – a Technical Memorandum on Reclamation and Disposal 1976
10	Local Authority Waste Disposal Statistics 1974/75; 1976
11	Metal Finishing Wastes – a Technical Memorandum on Arisings, Treatment and Disposal 1976
12	Mercury Bearing Wastes – a Technical Memorandum on Storage, Handling, Treatment Disposal and Recovery; 1977
13	Tarry and Distillation Wastes and other Chemical Based Wastes – a Technical Memorandum on Arisings, Treatment and Disposal; 1977
14	Solvent Wastes (excluding Halogenated Hydrocarbons) – a Technical Memorandum on Reclamation and Disposal; 1977
15	Halogenated Organic Wastes – a Technical Memorandum on Arisings, Treatment and Disposal; 1978
16	Wood Preserving Wastes – a Technical Memorandum on Arisings, Treatment and Disposal; 1980
17	Wastes from Tanning, Leather Dressing and Fellmongering – a Technical Memorandum on Recovery, Treatment and Disposal; 1978
18	Asbestos Wastes – a Technical Memorandum on Arisings and Disposal; 1979
19	Wastes from the Manufacture of Pharmaceuticals, Toiletries and Cosmetics – a Technical Memorandum on Arisings, Treatment and Disposal; 1978
20	Arsenic Bearing Wastes – a Technical Memorandum on Recovery, Treatment and Disposal; 1980
21	Pesticide Wastes – a Technical Memorandum on Arisings and Disposal; 1980
22	Local Authority Waste Disposal Statistics 1974/75 to 1977/78; 1979
23	Special Wastes – a Technical Memorandum Providing Guidance on their Definition; 1981
24	Cadmium Bearing Wastes – a Technical Memorandum on Arisings, Treatment and Disposal; 1984
25	Clinical Wastes – a Technical Memorandum on Arisings, Treatment and Disposal; 1984
26	Landfilling Wastes – a Technical Memorandum on ; 1986
27	Landfill Gas – a Technical Memorandum providing guidance on the monitoring and control of landfill gas; 1989, 2nd edition 1991
28	Recycling – a Memorandum providing guidance to local authorities on recycling; 1991

9.3 PRINCIPAL ACTIVITIES AND MAIN ENVIRONMENTAL EFFECTS

9.3.1 The scoping exercise

All aspects outlined in the Environmental Assessment Regulations must be examined at some stage. Some issues may be more sensitive than others and require more detailed examination. The selection of issues for more detailed examination is determined by a scoping exercise, which can also be used to identify the preferred site, should more than one option be available.

The scoping exercise comprises a combination of site visits and collecting initial data that are readily available (i.e. published data) and initial conversations with interested parties, including local planning authority and statutory consultees. The role of consultation with various parties, including the general public, is described in Section 9.3.5. Through this process it is possible to identify areas that are sensitive both geographically and environmentally and to highlight areas that may be subject to opposition from the local population. However, notwithstanding the interests of local pressure groups, further investigation work should be directed towards environmental characteristics that may be significantly affected by the development.

The scoping exercise should involve the use of specialists in the compilation and assessment of data. It should include an assessment of the construction and operational effects of the project against the current background environment and the environment that will pertain at the time of construction and operation (see also Chapter 2).

9.3.2 Baseline data requirements: project activities

It is important that environmental aspects are considered as early as possible in the development of any project. At conceptual design stage it may be that there are several alternatives being considered and environmental effects should be taken into account at this very early stage. For example, there may be several possible locations for a proposed incinerator development. A sensitive area, for instance with a school downwind of the proposed plant, could thus be eliminated from development in favour of a site where there are industrial areas downwind of the plant. However, the same kind of process design at each location could result in very different environmental effects, and it may also be possible to consider modifying the design to eliminate or minimise some of the adverse effects. In contrast, when assessing a proposal for a landfill site there may be only one location that is feasible, in which case the consideration of options will concentrate on engineering design possibilities and on the associated degree of security and risk of environmental impairment. Care should be taken to ensure that the degree of professionalism given to consideration of environmental effects is in no way limited because the choice of site is restricted.

Project activities include:

- site selection works including site investigation
- site clearance works including access and provision of services
- construction works and commissioning
- operation and maintenance of the facility
- any secondary development arising from the project
- decommissioning and demolition of the facility at the end of its useful life.

Baseline data must therefore be adequate to assess the effects of all these activities.

Waste facilities are generally divided into two types of category. The first is the transfer/ treatment station or incinerator plant where a process is taking place on the site. This generally involves a large building, the provision of mechanical plant and the provision of services as the major part of the development. Landfill, by contrast, is similar to a large earthworks project, generally with a lesser requirement for infrastructure. It can be seen that these two types of development have very different environmental effects and they are described separately below.

Environmental effects arising from the various phases (i.e. construction, operation and post-operation) of the development will be very different. In many developments the construction phase is relatively short but causes the most major disruption to local inhabitants. The operational phase, which may go on for the foreseeable future, may be perceived to be more acceptable environmentally. Thus although the provision of a waste facility may be beneficial to the community as a whole, proposals for its development will result in fierce opposition from the local residents. In addition, landfill sites resemble construction sites for the whole of their working life. Heavy earth-moving equipment and possibly noisy plant will be operating on the site from the first stages of site preparation to the final stages of site restoration. Added to this is the perception of nuisance, albeit easily mitigated, related to flies, pests and odour at landfill sites, and toxic fumes at treatment and incinerator sites. Aspects such as the possibility of contaminating groundwater that is used for local water supplies and the potential for landfill gas to migrate from the site and cause an explosion hazard are well documented and used by press and local pressure groups in opposition to landfill sites. This opposition may be offset in part by giving details of the mitigation measures to be employed and of the beneficial aspects of the proposal such as the use of methane as an energy source at a landfill site or heat/energy recovery from incineration.

9.3.3 Common environmental impacts associated with waste management

Table 9.5 lists activities that may result in potentially significant impacts. Note that the table shows the frequency with which such effects may be encountered and not the significance of the effect, as this is site-specific. Table 9.6 demonstrates the environmental characteristics for which the baseline data may be required when assessing waste management facilities. As stated previously, the different types of waste facility will have different types of effect on the local and regional environment and beyond. Specific effects are discussed under Section 9.4; however, general effects related to specific types of facility are outlined below.

Landfill sites

Two main environmental effects related to a proposed landfill development are the possibility of contaminating groundwater and the possibility of landfill gas migrating from the site and collecting in hazardous concentrations in nearby buildings. Another major issue is that of traffic congestion of local road networks. Historically, landfill sites have often been in remote locations with narrow winding lanes leading to them, and there is naturally a concern by the local population that large lorries (whether bringing clay for the liner or the waste for disposal) going through these lanes will cause disruption to the network and danger to children. Even where landfills are located close to major roads, the issue of increased numbers of heavy goods vehicles on the roads, particularly at sensitive junctions, still has to be addressed. Noise, general nuisance (seagulls – particularly near to airports – rats, flies and odour) and visual intrusion may also be significant effects in some locations.

It should be noted that exhausted mineral quarries are often colonised by species that may be rare in a particular locality; or there may be a rare outcrop of a particular sequence of geological strata that is of interest. Such sites may therefore be designated as Sites of Special Scientific Interest (SSSI) and, as described later, may necessarily be destroyed by landfilling activity at the site. Sensitive design of the facility may occasionally permit the SSSI to remain undisturbed, or allow habitats to be recreated once filling has been completed.

It may be necessary in certain locations and for certain facilities to carry out modelling of, for instance, groundwater from a landfill site to demonstrate that nearby abstraction points will not be affected by the proposed development.

Incinerators, toxic waste treatment plant

The main environmental effects associated with incinerators and toxic waste treatment plants are air pollution, residue disposal, visual intrusion and traffic.

Table 9.5 Activities that may result in potentially significant impacts

Phase/activity	Examples	Treatment	Disposal
Construction phase			
Site works	Land clearance; earthworks	■	■
Access	Earthworks	□	□
Infrastructure	Water supply, sewers, power supplies, communications, lighting	■	■
Raw materials demand	Lining materials, cover and capping		■
Transport of raw materials	Lining materials, cover and capping		■
Transport of employees			
Immigration	Contractors, temporary workforce	□	
Employment	Employment of local people		
Local expenditure	By developer on services and materials; employees on goods		
Lighting	Construction site		
Noise	Transport, pile driving, earthworks	□	□
Odours			□
Dust and particulates	Earth moving		□
Gaseous emissions	Exhaust emission		
Aqueous discharges	Pollution incidents, collision	□	□
Material disposal	Surplus overburden		□
Solid waste disposal	Materials on site, old structures, contaminated land	■	□
Accidents/hazards			□
Operational phase			
Structures	Buildings, weighbridges, wheelwashers	■	□
Raw materials demand	Capping materials		■
Transport of raw materials	Capping materials		■
Import of waste for disposal		■	■
Transport of employees			
Immigration	Permanent workforce		
Employment	Long-term job creation/loss		
Local expenditure			
Lighting			
Vibration	Traffic		□
Noise	Traffic, site plant	□	□
Odours		□	■
Dust and particulates			
Ongoing site preparation	As for Construction above		
Gaseous emissions	Exhaust emissions, stack emission, landfill gas	■	□
Aqueous discharges	Surface runoff, leachate, treatment liquor	□	□
Accidents/hazards	Collision, fire, explosion, structural failure	□	□
Post-operation			
Restoration	As for Construction above		
Cessation of activity	Employment, transport		
Deterioration of structures	Degradation, collapse	■	
Removal of structures	Demolition	■	□
Contamination	Long-term pollution of soil and water	□	■
Disposal of residues	Contaminated land	□	
Maintenance of landform	Landfill site		□
Aqueous discharges	Surface runoff, leachate		■
Gaseous discharges	Landfill gas		■
Accidents/hazards	Fire, explosion		■

Key:
■ Commonly causes potentially significant impact
□ Occasionally causes potentially significant impact

Note: This list is indicative only and does not necessarily cover specific sites.

Table 9.6 Baseline data requirements

Environmental characteristics	Examples of appropriate data
Flora and fauna	Habitat type; rarity; diversity; abundance; local/national significance; location of designated sites; specific characteristics of site which create habitat
Birds	Migratory birds; overwintering birds; nesting sites; protected species; feeding/loafing of gulls
Geology	Permeability; fissures; protected sites/exposures
Hydrogeology	Depth to aquifers; use of aquifers; water quality; protected zones
Hydrology	Drainage; surface runoff; channel characteristics
Surface water quality	Quality; status (classification); NRA objectives; trends; sources of pollution; data in accordance with relevant EC Directive(s)
Soil/ground conditions	Structure; bearing capacity; chemical composition; recompacted permeability and strength; depth of topsoil/subsoil; MAFF classification; contaminated land.
Air quality	Carbon monoxide, carbon dioxide, methane oxides of sulphur, oxides of nitrogen, metals, chlorinated compounds, radon
Climate	Rainfall; temperature; unusual characteristics, wind direction and strength
Population	Density; numbers; demographic features; industry, commerce and residential centres; trends in waste production, minimisation, re-use and recycling
Local community	Cultural status; interest groups; airports
Health and safety	Site records for site extensions; special characteristics; epidemiological data on cancers and respiratory illnesses
Emergency services	Site records for site extensions
Risk	Existing sources of risk; characteristics; trends
Land use	Urban; agriculture; derelict mineral workings; 'green-field' site; designated areas on development plans; proximity and type of nearest buildings
Landscape	Designated areas; topography; aesthetics; restoration, contouring, landforming
Leisure and amenity	Seasonal use of site; public rights of way
Tourism	Tourist attraction; number of visitors
Heritage	Archaeology; architecture; ancient monuments; listed buildings
Infrastructure	Roads; power; water; sewers; waste disposal; communications
Traffic	Road; rail; congestion; black spots; schools; status in relation to capacity; sight lines; junctions; bridges
Noise	Background levels

Incinerators, particularly municipal waste incinerators, are generally housed in very large buildings: hence although environmental effects during construction will be similar to those for any infrastructure development, they will go on for a considerable period of time. The need for particularly secure foundations (owing to the sensitivity of the structure to movement) may lead to special treatment techniques to improve the structural integrity of the ground, and to the need for massive foundations requiring extensive earthworks and significant quantities of concrete being placed.

Incomplete combustion can result in the emission of harmful gases and particulates, which may be implicated in high incidences of certain diseases in the vicinity of the incinerator. Poor combustion also results in waste which, following disposal, has a greater potential for pollution through the generation of gas and leachate. However, good design and operation will eliminate these problems. Visual intrusion arises from the presence of a stack/chimney and, if combustion control is poor, the presence of a visual plume emanating from the stack.

The treatment of toxic wastes may result in the generation of harmful gases although these are unlikely to be present at significant concentrations outside the plant. Poor control of the treatment process may result in contaminants being unexpectedly available for leaching following disposal of, say, treated sludges to landfill. Again, good design and operation can eliminate these problems.

In both cases traffic associated with the plant may be subject to local opposition but these facilities are generally proposed in industrial urban environments where the impact may be less severe than in a rural environment.

Modelling of air pollution emissions from an incinerator may be required in certain cases to show that sensitive local communities will not be affected.

Transfer stations and recycling centres

The main perceived effects arising from transfer and treatment stations and recycling centres are traffic issues, noise, odour and general nuisance. Traffic may be a major issue at recycling centres: some of these facilities receive over a thousand private vehicles per day. Such sites are generally located near to centres of population so that the public can bring in recyclables and bulky household items.

9.3.4 Baseline data requirements: environmental parameters

Environmental parameters that may be affected to a greater or lesser extent by waste management schemes are listed in Table 9.6 along with a checklist of common information requirements for the preparation of an adequate baseline data set. It should be noted, however, that this checklist is not definitive. In addition to collecting existing published and other data it may be necessary to carry out some field surveys and sampling to augment this data, particularly in the following areas.

Physical and chemical characteristics

- *Climate and ambient air quality*: Particularly important in the case of a proposed incinerator development.

- *Geology*: The permeability/porosity of the geological strata around and underneath the proposed development site; presence of fissures, faults or other discontinuities, whether the site is a geological or geomorphic SSSI.

- *Groundwater*: Depths to aquifer, use of aquifer; whether this is close to the site or distant from the site and quality of water within the aquifer; continuity between aquifers.

- *Surface water*: The characteristics and sensitivity of local streams and rivers, particularly if there are nature conservation interests or fisheries.

Biological characteristics

- *Ecology*: Site surveys to gather ecological information and, if the site has an SSSI designation, then to establish the extent of the SSSI and the reason for the designation; also to establish the sensitivity of particular habitats and their life to different aspects of the development proposal.

Human characteristics

- *Traffic counts*: To establish the capacity of the local network and the traffic currently using that network; to assess both construction effects and operational effects.

It may frequently be necessary to carry out long-term monitoring prior to making a judgement on the significance of any environmental effect. Such would be the case with, say, air quality monitoring, groundwater quality and noise. Ecological data may have to be collected over an equally long period of time to ensure that, say, wet and dry seasons are covered, breeding seasons are monitored and migration patterns can be established. Natural variability in biological communities with time should be assessed.

9.3.5 The role of consultation in the data-collection process

While much of the information for an EA may already exist, it may not be in published form but retained in the memories of people who live around the site, local nature conservation groups, or local authority personnel who had cause to deal with the site in the past. It is therefore important where possible to make contact with such personnel and to obtain the information from them. Early discussions with the National Rivers Authority (NRA) for instance will indicate immediately the potential sensitivity of an aquifer underlying a site and officers will be able to give an indication of the licensed abstractions in the area that may be affected. This information is useful to augment that obtained from the registers that are available to the public. It should be noted that the NRA is a statutory consultee under certain applications for waste disposal facilities, and controls aqueous discharges direct to controlled waters as well as the environmental effects of the development.

It is advisable to talk to the planning authorities and statutory consultees at as early a stage as possible to find out their concerns about the facility and the sensitivity of the area, as this will affect the direction of the EA. Statutory and other consultees include English Nature, Countryside Council of Wales, English Heritage, Countryside Commission, Historical Buildings and Monuments Commission and the National Rivers Authority. The roles of the various agencies controlling waste management are given in Section 9.2.3 (see also Table 2.2).

While consultation is a valuable tool, it should be used extremely carefully as it is very easy to raise public alarm by consulting with interested parties at too early a stage in the development process. These groups would be interested to hear what mitigation measures are to be incorporated in the design, and if such measures have not yet been designed, the proposals may only cause alarm. Consultation with local authority officers should in any case identify areas of concern to the local public. However, as development proceeds to a more final form, it may be useful, depending on the significance of the development to the local area, to hold some sort of consultation meeting with the general public to inform them and to alleviate any fears (see also Section 2.5.2).

With waste management facilities, continuous liaison with local groups is necessary to ensure good relations and to ensure that operational changes are amenable to all parties. Unlike other types of development, the NIMBY syndrome perpetuates throughout the life of the site.

Box 9.3 Case study: The Welbeck Project

The Welbeck reclamation and landfill project is one where public consultation has been used successfully.

The project involved the diversion of the River Calder to permit an area of derelict land, with no amenity value, to be used for the disposal of municipal wastes and colliery spoil.

Advance publicity was given at significant stages in the planning process and included coverage in the local media. An exhibition and model were prepared to encourage public dialogue. This has continued with the issue of **Briefing Notes** on particular aspects and an open day to which interested parties were invited. An Interpretation Officer has now been appointed to provide the public with information on the whole scheme and a purpose-built visitor centre will soon be opened.

9.4 IMPACT IDENTIFICATION, EVALUATION AND MITIGATION

9.4.1 General

This section deals in broad terms with the identification, prediction, assessment and mitigation of possible impacts. Methods for assessing the significance of impacts, particularly against any existing or potential environmental quality standards or objectives, are given in Tables 9.8−9.11 and sensitive resources are highlighted. Suggestions are put forward for mitigating measures to alleviate adverse impacts and to enhance the environment. Sections 9.4.2−9.4.4 deal respectively with physical and chemical, biological and human environmental characteristics. As the adverse or beneficial nature of potential impacts will to a large extent be site specific, no reference is made to this in the tables.

The physical and chemical characteristics of the existing environment listed in Tables 9.5 and 9.6 can potentially be affected in a number of ways by waste management schemes (Table 9.7).

Table 9.7 Summary of the impacts considered in this chapter

Impact	Section no.	Table no.
Air quality	9.4.2	9.8
Water quality and groundwater	9.4.3	9.9
Land: soil and geology	9.4.4	
Flora and fauna	9.4.5	9.10
Socio-economic/cultural	9.4.6	
Leisure and amenity	9.4.7	
Land use and landscape	9.4.8	9.11
Health and safety	9.4.9	
Traffic	9.4.10	
Heritage and archaeology	9.4.11	
Noise	9.4.12	

9.4.2 Air quality

Many waste management schemes give rise to effects on air quality both during the construction and operational phases. Table 9.8 gives examples. During the construction phase of the works — whether for incineration or treatment works or a landfill site — there may be some local deterioration of air quality arising from site clearance works. This can be mitigated by damping down materials prior to demolition or soil stripping. Where large quantities of materials are brought in by road — for instance clay for provision of a liner in a landfill site — there may also be local deterioration in air quality arising from exhaust emissions. However, the greatest potential for deterioration in air quality arises during the operational phase of landfill sites, treatment works or incinerators. At all three types of facility there may continue to be local deterioration of air quality from vehicles arriving at the plant bringing waste materials for treatment or disposal. In addition, at a treatment plant there may be chemicals being brought to the site; and cover materials may have to brought in bulk to landfill sites.

Following placement of waste in a landfill site, gases are produced whose trace components can cause an odour nuisance to nearby residents. The major components of landfill gas (i.e. methane and carbon dioxide) can pose a hazard to nearby buildings and their inhabitants. Incineration plants generate stack emissions which, if not properly controlled, can result in nuisance to the nearby residents. Emissions arising from the thermal treatment of particular wastes (e.g. PCBs) can also, if not properly controlled, generate particularly toxic emissions (e.g. dibenzofurans and dioxins). However, there are strict controls on these type of processes, which should ensure that such hazards do not arise.

While it is difficult to mitigate emissions from moving vehicles except through the use of fewer numbers of larger vehicles, all other emissions arising from the operation of waste facilities can be satisfactorily controlled. For instance, at a landfill site the use of bunds to screen the site from downwind receptors, immediate covering of placed waste and the extraction and flaring of all gases can effectively eliminate an odour problem. Masking sprays (while in themselves not particularly efficient in reducing odour problems) can be used to provide psychological assurance that the problem is being addressed. Collection and flaring of all landfill gases, combined with a robust liner or cut-off system around the site, can significantly reduce the hazard to properties close to the site. Should it prove necessary, further passive venting can be installed between the site perimeter cut-off system and the properties.

Mitigation of stack emissions from an incinerator can be effected by the installation of gas-cleaning equipment and particulate precipitation and collection, by altering the stack height to provide improved dispersion and by control of the grate temperature and residence time. Control of gaseous emissions from chemical and other waste treatment plant can be effected by providing negative pressure at all possible sources and collecting all (compatible) gases together for treatment prior to release. Dust emissions from silos can be treated by providing negative pressure, by enclosure of the silos and by damping down areas where emissions cannot be prevented.

Table 9.8 Examples of the impact of waste management schemes on air quality

Issue	Possible cause	Typical effects	Predictive techniques	Mitigation and enhancement options
Chronic air quality deterioration	Vehicular emissions	Nuisance, chronic health problems	Baseline assessment	Use bulk haul transfer
Odour nuisance (landfill)	Trace elements in landfill gas Disposal of putrescent or odorous wastes Incomplete daily cover	Nuisance to local residents, effect on local amenity	Baseline assessment	Bunds to protect downwind receptors Daily cover/improving daily cover Masking sprays Gas extraction and flaring/utilisation
Odour nuisance (incinerator)	Incomplete combustion Waste storage	Nuisance to local residents, effect on local amenity	Baseline assessment, modelling, monitoring	Improve combustion e.g. raise temperature, increase retention time; install scrubbing equipment; improve waste throughput
Particulates	Dust from earthworks and landfill operations	Local deposition as nuisance, visual quality	Baseline assessment, monitoring	Use of spray equipment, foams
	Stack emissions	Local deposition as nuisance, visual quality	Baseline assessment, modelling, monitoring	Improve combustion control, install scrubber, precipitation equipment
Hazard to nearby buildings and inhabitants	Landfill gas migration	Asphyxiation Explosion Injury, loss of life	Monitoring in waste, ground and buildings	Install cut-off / liner around site Collect gas and flare Install passive venting as secondary protection

9.4.3 Water: water quality and groundwater

Most waste materials will degrade to produce water-soluble or water-transportable products which (as leachate) can contaminate both ground and surface waters. The pollution of groundwaters is particularly important when aquifers are used for abstraction, especially for potable or industrial use. The EC Groundwater Directive, which has been in force for some 10 years, and now the National Rivers Authority's groundwater protection policy require protection of all aquifers, regardless of their use now or in the future. It is therefore important to assess the potential for contaminating any groundwater body.

Surface waters can be polluted by the unintentional discharge of leachate with either ground or surface water (e.g. where groundwaters have been intercepted, routed around or through a site and then discharged to a stream). Also, the build-up of leachate within a landfill site may result in an outbreak of leachate which then runs down into, say, a stream. The most likely effect is not destruction of habitat but of fish being killed either through the introduction of high concentrations of ammonia which few fish can tolerate or through reduction of oxygen in the water. Invertebrate life may be similarly destroyed. If the leachate has arisen from a co-disposal site (i.e. one taking toxic or special wastes) there may be chemicals within the leachate which themselves can cause harm to fish.

Table 9.9 Examples of the impact of waste management schemes on water resources

Issue	Possible cause	Typical effects	Predictive techniques	Mitigation and enhancement options
Chronic groundwater pollution	Dilute and disperse landfill site	Aquifer contamination	Computer modelling sampling/monitoring	Collect, pump and treat leachate where possible
Acute groundwater pollution	Breakdown in landfill liner containment system	Aquifer and contamination	Computer modelling sampling/monitoring	As above
Acute surface water pollution	Leachate outbreak	Acute fish kill	Monitor leachate levels in site	As above. Ensure integrity of above ground containment systems: i.e. bunds and runoff collection systems
Chronic surface water pollution	Leachate contaminating 'clean' discharge	Chronic fish kill; growth of 'sewage fungus'	Monitoring	Trace and isolate source of pollution; collect and treat discharge
Chronic surface water pollution	Recharge by contaminated groundwater	Chronic fish kill; growth of 'sewage fungus'	Monitoring	Trace and isolate source of pollution; prevent recharge from this source if cannot be treated
Aquifer flow	Plug of low-permeability or impermeable materials introduced into aquifer	Reformation of flow lines	Modelling	Amend shape of landfill site to reduce
Aquifer recharge	Area of low-permeability material (or impermeable) being placed on recharge area	Shadow on recharge, reduced recharge	Modelling	Amend shape of landfill; collect clean runoff for use in recharge

A further aspect to be considered is that of surface water recharge by groundwater which itself has been contaminated by leachate. Table 9.9 outlines a number of surface and groundwater quality issues that might result from waste management schemes.

A further major issue related to landfill is the loss of recharge to an aquifer resulting from the placing of a low-permeability or even impermeable mass over an area where previously there were permeable materials. Also, for infilled quarries, the effect of placing a plug of low-permeability material within a high-permeability stratum must be considered.

In some cases, waste disposal is used as a means of reclaiming marginal or derelict land; where such land is a wetland there are separate issues to be considered, for instance, loss of habitat.

The solid residues from the physical or chemical treatment of waste and incineration are generally sent to landfill for disposal, so the effects are as outlined above for landfill. Note that this includes the dust and ashes precipitated within the incinerator chimney, and although the process of treatment or incineration will have concentrated contaminants (principally heavy metals, the volatiles having been burnt off) within the sludge and ash for disposal, the total volumes are much less than if the material had been directly disposed of. Aqueous discharges may in some cases be made to controlled waters. In these cases the National Rivers Authority will specify its requirements to protect the water environment.

Groundwater may be protected from contamination by leachate through the installation of a landfill liner. The degree of protection is dependent on the sophistication and robustness of the liner design. Previously it has been acceptable to install a clay liner to no particular standard provided the clay was compacted. Today 1 m of clay compacted to give a permeability of 1×10^{-9} m/s or less (or the equivalent of this) would be the minimum acceptable. However, the NRA are generally requiring an improvement on this in the provision of double or triple liner systems with built in leak detection and the draft EC Directive on the landfilling of waste requires the equivalent of 3 m of 1×10^{-9} m/s. Synthetic membranes provide virtually nil permeability whilst they remain intact, but their long-term robustness and integrity still requires verification.

Separation of ground and surface waters, together with the installation of liners or cut-offs to separate clean and contaminated waters, will eliminate the accidental discharge of contaminants into surface water bodies. The provision of secure above-ground containment systems, together with the management of leachate in such a way that levels do not build up within the site, will eliminate the possibility of leachate outbreaks. For sites located close to particularly sensitive surface water bodies, it may be necessary to give fail-safe protection through the provision of a surface runoff collection system, passing through a holding lagoon whose discharge is controlled through valves.

Much of the discussion in this section has been given over to the effects arising from landfill. Transfer stations, incinerators and treatment plants may have significant effect on water resources, particularly if they involve the development of a contaminated land site, the storage of hazardous materials or aqueous discharges to controlled waters. For example, the proposed Doncaster incinerator was refused planning permission because of the possible impact on an aquifer.

9.4.4 Land: soil and geology

The primary effects of waste management schemes on soil and geological interests arise from landfilling activities. As many landfills are proposed for the restoration of worked-out mineral quarries, consideration has to be given to the ecological value of any habitat which may have become established in the quarry, and to the exposure of particular geological sequences which may be of interest. Parts or even all of the site may have been designated a Site of Special Scientific Interest or the regional or local equivalent for these reasons. While it may be possible to design a landfill site that accommodates a particular habitat or permits the continued exposure of a particular geological sequence, in most cases proposals for the landfill site will necessarily result in destruction of sensitive habitats on quarry floors and the covering up of geological sequences if these are at low level in the quarry.

The practice of above-ground landfilling or landform is becoming more common. This generally takes place on low-grade agricultural land, and consideration must be given at an early stage to the restoration proposals, which may include returning to the same grade of agricultural land.

The construction of incinerators or waste chemical treatment plants will have the same potential disturbance of the ground as any other building. However, these facilities handle hazardous and toxic materials which produce heavily polluting liquors. Spillage of waste materials or the liquors, where the ground is unprotected, can lead to contamination of the ground and groundwaters. Common practice is to seal the surface of all areas where wastes may be exposed and to provide drainage facilities for the collection of liquors. The effects of construction and operation will therefore be site specific, and only of real significance in particularly sensitive locations. With regard to the disposal of treated wastes or incinerator ash, the effects are the same as for landfill.

Proposals for the aftercare of landfill or landform sites should allow for settlement. The biodegradation of the organic contents of the landfill will inevitably result in settlement at the surface. The degree of settlement and the variation across the site can be minimised both through good engineering design (i.e. structural integrity of the foundation of the site) and through good operational practice (i.e. thin layers, well compacted, leachate management). The stability of the site, or parts of the site (bunds, tipping faces) can be similarly ensured.

9.4.5 Flora and fauna

The biological communities in the area (i.e. not just the site alone) which may be affected by the development proposal need to be defined in order to predict likely effects, to identify monitoring requirements and to determine what remedial action may be required.

Sites selected for potential landfilling have often lain derelict for some considerable period of time. (This is, of course, not the case where reclamation through infilling with waste is proposed as part of a planning application for mineral extraction and the landfilling follows on immediately from the mineral extraction.) Derelict sites, including worked-out quarries, are naturally vegetated after a period of time with a succession of species which can be unique in a particular area. In some cases the particular conditions on that site may attract certain species that are rare nationally.

For instance, areas of chalk downland have had the chalk excavated for cement manufacture with excavation taking place down to the clay. This has resulted in ponds being formed where species such as Great Crested Newts have colonised. In other areas, gravel pits have become filled with water owing to the local high water table and have become colonised by rare birds and used by migrating birds, resulting in the sites' designation as SSSIs. The loss of these derelict areas therefore becomes significant. The disposal of, for example, pulverised fuel ash into coastal lagoons may result in the loss of wading areas. There is also the potential problem of a plume of suspended solids spreading out into the sea and affecting spawning grounds.

In such cases it is important to discuss with English Nature the precise importance of the site, both nationally and locally. It may be possible (e.g. in the case of Great Crested Newts) to collect the protected species and to translocate them, followed by the further possibility of re-introducing them to the site following completion and restoration (for a landfill), given that a suitable habitat can be provided. It may be possible in a landfill restoration phase to provide or extend a unique habitat that attracts special nature conservation interest. Sediment release into coastal waters can be controlled by suitable design of the retaining wall.

Box 9.4 Case study: Witton Landfill

At the Witton landfill site, Northwick, an innovative restoration scheme was established in which a conservation area of transplanted vegetation has been combined with existing woodlands, hedge and pasture. In land adjacent to the landfill sites, a series of limebeds containing rare and interesting species were at risk due to invasion by birch and hawthorn. An assessment was undertaken that considered the feasibility of saving the limebed flora by transplanting turves from it onto a completed part of the landfill site. Transplantation experiments were conducted, after which a scheme was implemented. When transplantation is complete, the local council, in conjunction with the Nature Conservancy Council (English Nature), aim to declare the whole of the Witton site a statutory nature reserve to ensure the viability of the newly formed conservation area.

With regard to incineration there is the possibility that emissions from the incinerator stack may be deposited over a large area downwind of the site, causing degradation in the quality of habitat and perhaps affecting certain species of flora and fauna. Direct deposition of particulates and acid gases will affect fish as will the uptake into rain and subsequent deposition. Effects relating to the rate of incidence of certain diseases in humans are discussed under Section 9.3.3. Mitigation methods are those required to control gaseous and particulate emissions and are described under Section 9.4.2.

As already mentioned above, the constituents of leachate have a high potential for affecting fish. Most fish are acutely sensitive to elevated ammonia concentrations and to the reduction in oxygen levels which may result in their water body following a discharge of leachate. Specific heavy metals and other constituents of the leachate, particularly organic compounds, are toxic to certain species of fish. Other water life, such as invertebrates on which fish feed, can also be affected. Some contaminants become bound up with the sediments in a stream, pond or river and can cause long-term damage as they diffuse back out of the sediments once the water itself has become clean. A similar chronic effect occurs when polluted groundwater provides recharge to streams and ponds. Estuarine and coastal waters may also be affected in this way. Mitigation methods are those required to prevent contamination of ground or surface waters and are described under Section 9.4.3.

Landfill gas (comprising methane and carbon dioxide) is an asphyxiant. Methane in particular can collect around the roots of plants, starving them of oxygen and causing dieback. This is particularly noticeable along fissures in the surface of a landfill and in the condition of trees planted near to landfill sites. However, appropriate techniques (e.g. gas collection and venting or flaring, low-permeability capping, increased depth of soil cover in tree planted areas) can ensure that no die-back occurs either in trees or in smaller plants. Under the right conditions, i.e. where there is pressure to force methane into solution in water, the gas can then be carried in the water and released at a later stage. This will result in anoxic conditions at the point of release. There have been a number of cases of claims for compensation of crop damage arising from methane migrating or being carried out of landfill sites and later released and subsequently surfacing through fields planted with crops. It is further considered that it is possible for methane to be generated at a point remote from the site to which leachate has migrated and where conditions may be suitable for anaerobic degradation to take place. Mitigation methods are those required to control the hazard of landfill gas and are described under Section 9.4.2.

Further details are given in Table 9.10. Pre- and post-scheme biological survey work is often likely to be required to ensure protection of the whole ecosystem.

Table 9.10 Examples of the impact of waste management schemes on flora and fauna

Issue	Possible cause	Typical effects	Predictive techniques	Mitigation and enhancement options
Destruction of habitat	Site clearance; drainage, diversion of water courses, landfilling, pipeline construction	Loss of terrestrial or aquatic habitat, loss of communities	Baseline assessment	Select options and carry out operation to minimise environmental disturbance; recreate habitat on adjacent site or on development site following cessation of works
Pollution of streams etc.	Leachate outbreak or seepage	Fish kill	Monitor effect: survey Baseline assessment	Collect pump and treat leachate
	Spillage of waste or chemicals	Loss of communities, damage to habitat	Monitor effect: survey Baseline assessment	Removal of spilt substance from water by pumping etc.
	Insufficiently tight consent conditions	Damage to biological communities	Baseline assessment, followed by monitoring	Tighten discharge consent conditions
Deposition of airborne pollutants	Particulate or gaseous stack emissions from incinerator	Damage to habitat; loss of community, fish kill	Baseline assessment, computer modelling	Improve combustion control Install scrubbing equipment
Crop damage	Gas migration	Crop die-back	Monitoring	Gas extracting and flaring; install cut-off Passive venting

Note: Baseline assessment — would be made before the scheme became operational.
If pollution occurred, its effect would be monitored using data upstream/downstream of the site, where possible, plus the baseline assessment.

9.4.6 Socio-economic/cultural

Most waste management schemes will have only a limited impact on socio-economic and cultural parameters. The economic significance is more evident in the need for effective waste disposal provision to support local economic activity than in employment opportunities. Employment opportunities which are created as a result of construction processes are likely to be relatively short-term and, for landfill, to involve relatively few people. During the operation phase of chemical treatment plants and incineration plants, again relatively small numbers of people are likely to be employed and although these may be semi-skilled it is likely that they will come from the local area. Landfill operations will tend to employ people who may be resident in the area and who have skills of manoeuvring earth-moving plant, of labouring, or technician-level monitoring skills. Following the implementation of the waste-licensing provisions of the Environmental Protection Act 1990, all persons operating waste facilities will be required to demonstrate that (managerially or technically, as appropriate) they are a 'fit and proper person' under the Act. To do this, they must obtain a certificate on the basis of having achieved national vocational qualifications. Landfill may sterilise the land for certain post-operation phases for a considerable period of time. While it is possible that the reclamation of the landfill site may result in a use that generates employment for the local community, e.g. the construction of a super store, it is more likely that the end use would be amenity or agricultural and therefore likely to employ very few people. There are unlikely to be any significant knock-on/follow-on effects resulting from employment issues related to these sites.

9.4.7 Leisure and amenity/tourism

Historically, many landfill sites have been restored to agriculture. However, with the downturn in the agricultural economy and the 'setting aside' of agricultural land, perhaps the most common form of reclamation for a completed landfill site today is for leisure and amenity. Many proposals for landfill include the construction of sports fields on the completed site. Increasing interest is shown in the development of nature parks and leisure parks on completed landfill sites. Proposers see this as a way of providing reassurance to consultees such as English Nature, and the public generally, of the long-term environmental benefits. This is also a way of returning a scarred industrial landscape to a use that is more acceptable environmentally. However, some schemes are over-ambitious, requiring the removal before landfill and the replacement after restoration of precious species and of the construction of ponds and water bodies on the top of landfill sites.

In some cases the landfill proposal will actually remove a leisure facility: for instance, where there is a proposal to infill a water-filled gravel pit where water skiing or sailing may be taking place at the current time. Other examples where landfill proposals may affect amenity and leisure are where disused quarries have become revegetated and are used by local inhabitants for walking their dogs and by local children for playing. It may be necessary to consider the effect on tourism if there is a particular geological sequence or special species in the area which attracts tourists; or if indeed the proposed facility (landfill, incinerator or chemical treatment works) interferes with or causes a nuisance to an established tourist attraction in the area.

9.4.8 Land use and landscape

The type of effect on land use and landscape will depend on the facility and its proposed location. Chemical treatment plants, incinerators and landfill sites will generally only receive planning permission if the location of the facility is in accordance with the local plan, i.e. industrial-type processes such as chemical treatment plants and incinerators may be restricted in location to areas designated for industrial use. Such a facility is then likely to be of a size and structure commensurate with its surroundings. The effect is therefore likely to be minimal. However, four toxic waste incinerator proposals have stalled at the planning stage because of extreme opposition for these facilities even in industrial areas. There is similar opposition to domestic waste incinerators.

An incinerator will invariably have a high chimney for dispersion of stack emissions. It will therefore be easily visible, probably at a considerable distance from the plant. It is likely that planning permission for an incinerator will only be granted in an industrial area and therefore the type of plant will itself probably be acceptable in the surroundings. However, the high visibility of the chimney may have a significant impact, which will be exacerbated if smoke or steam is seen coming out of the chimney.

Because landfill sites are often proposed as infills to worked-out mineral extractions, they are often in rural/agricultural areas. The Countryside Commission (1987) published guidelines on the assessment of landscape impacts that are particularly important in areas designated as being of special landscape interest. It is particularly important to take such issues into consideration when planning a landfill site both for the impact on the existing land use during filling and so that the restoration contours can fit in with the existing landscape. More details are given in Table 9.11.

Table 9.11 Examples of the impact of waste management schemes on land use and landscape

Issue	Possible cause	Typical effects	Predictive techniques	Mitigation and enhancement options
Construction site	Plant; stockpiling; structures; site offices	Visual impact; temporary loss of amenity/agricultural land; focus of interest for locals/visitors	Landscape assessments, Countryside Commission CCD (1987)	Minimise area effected; screening; reinstate and improve site after construction
Landfill site	As above	As above, but long-term effect; 'blot on the landscape'	As above	Screening, tree planting; careful selection of vegetation
Permanent visual intrusion	Structures; operations	As above	As above	As above
Permeability	Surfaces, slopes, structures	Increased surface runoff, loss of habitat, change of land use	Modelling	Collect runoff and re-introduce, recreate habitats

9.4.9 Health and safety; emergency services; risk

As with any civil engineering project, waste management schemes will potentially impact upon health and safety and risk characteristics during both the construction and the operational phases of the project. The degree of risk may in fact be exacerbated by the very nature of the waste materials: e.g. sharp corners, dangerous chemicals, explosive substances. A risk assessment combined with a COSHH assessment will indicate the probability of such an event but it will be necessary in most cases to have a well-rehearsed contingency plan in case of emergency.

There are particular risks associated with landfill sites arising from the degradation of the wastes deposited there. At certain concentrations landfill gas (comprising principally methane and carbon dioxide, with small quantities of trace gases) is both explosive and an asphyxiant. Methane has a lower explosive limit of 5% by volume in air and an upper explosive limit of 15% by volume in air. As landfill sites can generate up to 65% of methane by volume of the gas, there is thus a significant risk that methane can collect in enclosed spaces up to the explosive limit. The presence of an ignition source and sufficient oxygen will result in an explosion. It is therefore particularly important to have a contingency plan available for use when methane concentrations are nearing a specified level requiring evacuation. This is over and above any licence or other legal requirements for gas management systems. Carbon dioxide is present in normal atmospheres at around 0.3% volume in air. In buildings it is considered acceptable to have concentrations of up to 0.5% by volume in air for a working day, or for a very short period of time at 1.5%. As concentrations of carbon dioxide in landfill gas can reach as much as 35% it is therefore important that this gas is not permitted to accumulate in enclosed spaces where man entry may be required. Landfill gas may be generated, and hence pose a potential hazard, over several decades. Management for its control will be required until such time as the concentration of methane falls below a level where it poses a risk and remains below this level for a couple of years (specific guidance is given in WMP 27).

In addition to the specific hazards described above methane is an ozone depletant and contributes to global warming.

Box 9.5 Case study: Loscoe

> The most significant incidents arising from the collection of methane within enclosed spaces are the Loscoe disaster in Derbyshire in 1986 where methane collected in a bungalow and exploded, luckily with no loss of life; and the Abbeystead disaster in May 1984 where methane dissolved in water was released into a pumping station and the explosion resulted in loss of life. It should be noted that, in the second example, the source of methane was not a landfill site.

The hazard associated with landfill gas can be mitigated by the installation of a liner or cut-off around the site and the extraction and flaring of gas generated by the waste. A passive venting system can be installed outside the cut-off as a secondary defence against migration. Utilisation of the collected gas may prove a positive benefit to local industry. Both flaring and utilisation of the gas reduce the contribution to global warming.

Box 9.6 Case study: Stone landfill gas plant

> The Blue Circle Industries Stone Landfill Gas project at Stone, nr. Dartford, Kent, is the largest multi-user landfill gas facility outside the USA. The 40 ha, 10×10^6 m³ site produces 5,000 m³/h of gas (at 20°C/atmospheric pressure) which is supplied via a 7 km pipeline to cement kilns, lead smelting vessels, a whiting dryer and 1 MW electrical CHP system. The plant, operational since July 1986, also produces 2 MW electricity on site, utilising low methane landfill gas previously flared as part of the gas migration control system. Housing is located within 25 m of the site boundary on two sides, with effective gas control being maintained throughout.
>
>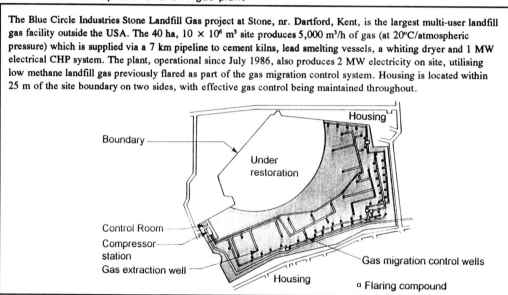

For chemical waste treatment plants there is a significant potential risk of spillage of chemicals arriving at the plant and of toxic materials also arriving at the plant for treatment. Issues of health and safety relate to handling chemicals and toxic wastes, to their mixing and the possibilities of synergistic reactions, and to the safe disposal of washings from treatment tanks.

For incinerators there are two main areas of health and safety related issues: the health and safety of workers in the plant handling materials for incineration and post-incineration; and the issues related to airborne emissions from the plant. In relation to the health and safety of workers, the risks described above for chemical treatment plants apply: i.e. the risk of spillage, of handling the materials, and of disposing of the residue. With regard to airborne emissions, the safety issues relate to combustion temperature and residence time to ensure complete combustion to a safe end product, and the use of efficient scrubbing systems to reduce particulate emissions and acid gas emissions from the stack.

Risk and hazard (HAZOPS) assessments of the operation of the proposed development can identify events with a high probability of occurring and of severe consequence and also identify measures that can be incorporated into the design or operation of the plant to reduce the probability or consequence of the event.

9.4.10 Traffic: transport and infrastructure

Both the construction and operation phases of waste management schemes will have the potential to disrupt access to local facilities and generate traffic. Depending on the type of facility and of the proposed throughput at the facility, it may be possible to use bulk transfer methods, i.e. barge or train for the transport both of materials for construction and of waste materials for treatment and disposal at the site. This will minimise the potential impact on the local transport facilities.

Because landfill sites have historically been in remote rural locations, often close to small village communities, the potential effects on local road networks are generally seen to be fairly significant. This may be the case even where the site is close to major roads. It is therefore not unknown for a developer of a large landfill site to offer minor road improvements such as better junctions or longer sight lines to improve the road safety related to increased numbers of heavy goods vehicles arriving at the facility. It may also be necessary to discuss with the local authority appropriate routes for vehicles using the site and ensure by the provision of signposting that these are used. Because landfill sites offer low employment opportunities in the locality and cause potentially significant effects on the road transport system together with environmental nuisance, the possibility of heavy goods vehicles passing along narrow lanes and small villages is often an emotive and important issue to the local inhabitants. Associated with this is the issue of noise, which is described later. The use of transfer stations to reduce the volume of waste for transport will similarly reduce the number of vehicle movements at the landfill sites which they serve. Rail haul transfer of waste direct to the landfill site will remove these road-haulage-associated effects.

Incinerator sites and chemical treatment sites will have more vehicle movements per tonne of waste as the waste is both arriving at the site and being taken away − possibly in different types of vehicle. However, as the facilities are normally located in industrial areas, the increase in goods vehicles may not be considered significant. Economies of scale require domestic waste incinerators to serve 200,000 to 500,000 people − this population may be spread over a large area requiring transfer to bulk haul transport. If rail haulage is practical, many of the effects on road traffic will be eliminated.

Traffic is a major issue at some recycling centres (see Section 9.3.3) with large numbers of private vehicles visiting the site each day. Situated, as they are, close to the public who wish to use them, these increases in vehicle movements have a significant effect on the roads leading to the site. Transfer stations are similarly generally situated fairly close to the centres of population, industry or commerce as their main function is to reduce the volume (by increasing the density) of wastes requiring transport to distant disposal sites. There may be substantial numbers of heavy goods vehicles (often large) arriving at and departing from such a facility, and the significance of the effect on the local road network will depend on the location of the site.

Traffic movements at recycling centres could be substantially reduced if recyclables were sorted at and collected from the household.

Depending on the location of the proposed development it may be necessary to construct special access facilities to minimise environmental effects (for instance, one landfill site required the construction of a bridge over a navigable river, another the construction of several miles of haul road over hilly terrain). Other infrastructure which may be required includes, *inter alia*, water supply, foul and surface water sewers, power supplies. It may be possible to supply the power requirements of the plant from treatment of the waste (e.g. waste to energy plants, utilisation of landfill gas) which will reduce environmental effects at local levels, national and beyond.

Box 9.7 Case study: Arpley bridge, rail terminal now being considered

Arpley is one of the largest above-ground landfill sites to be engineered in Britain. The site, which is operated by Cheshire County Council, is situated 4 km to the southwest of Warrington. Filling commenced in October 1988 after construction of site infrastructure. To enhance site access, two potential bridge routes were identified which crossed the River Mersey. The construction of the adopted new bridge not only enabled access to the landfill operation but also provided access for the opening up of additional industrial land.

9.4.11 Heritage and archaeology

Proposals for landfill sites, incineration plant and chemical treatment works generally do not impinge upon heritage and archaeology sites. However, this may not be the case with land form (i.e. above ground landfill). If the site is already listed or designated in some way, a variety of controls exist to protect the feature in question. However, if previously unknown features of interest are found part the way through the construction process considerable delays might be expected. The county archaeologists and various voluntary agencies should be able to advise at an early stage whether or not there is likely to be significant heritage interest at a particular site.

9.4.12 Noise

While there may be some considerable effect of noise arising from the construction of incineration plant and chemical treatments plants, it is unlikely that this will continue during the operation phase. The disruption effect is therefore minimal and short term. However, traffic and operations at incinerators continue 24 hours a day, seven days a week with the associated noise effects.

At landfill sites the construction-type activities continue for the duration of the landfill site's life, albeit for 12 or less hours a day Monday to Friday and Saturday mornings. Earth-moving equipment, bulldozers and compactors may operate almost continuously during the working hours of the site. If the landfill site is a former quarry it is probable that the noise perceived by local inhabitants will increase as the height of fill increases, becoming more obvious and more of a nuisance as the site nears completion. It is therefore important to erect bunds around the site and for other methods of noise mitigation to be employed; while there is noise associated with their construction, it is short lived and they shield the local inhabitants from further noise nuisance. Another aspect of noise related to landfill sites is the revving of engines, reverse warning tones, the tipping action of empty skips and rattling of chains in empty vehicles as the lorries drive up and down steep slopes into and out of the landfill site.

LEEDS COLLEGE OF BUILDING LIBRARY
NORTH STREET
LEEDS LS2 7QT
Tel. (0113) 222 6097 and 6098

16 OCT 2002